Springer Textbooks in Earth Sciences, Geography and Environment

The Springer Textbooks series publishes a broad portfolio of textbooks on Earth Sciences, Geography and Environmental Science. Springer textbooks provide comprehensive introductions as well as in-depth knowledge for advanced studies. A clear, reader-friendly layout and features such as end-of-chapter summaries, work examples, exercises, and glossaries help the reader to access the subject. Springer textbooks are essential for students, researchers and applied scientists.

More information about this series at http://www.springer.com/series/15201

Antonio Pulido-Bosch

Principles of Karst Hydrogeology

Conceptual Models, Time Series Analysis, Hydrogeochemistry and Groundwater Exploitation

 Springer

Antonio Pulido-Bosch
University of Granada
Granada, Spain

ISSN 2510-1307　　　　　　　ISSN 2510-1315　(electronic)
Springer Textbooks in Earth Sciences, Geography and Environment
ISBN 978-3-030-55372-2　　　　ISBN 978-3-030-55370-8　(eBook)
https://doi.org/10.1007/978-3-030-55370-8

This Springer imprint is published by the registered company Springer Nature Switzerland AG
The registered company address is: Gewerbestrasse 11, 6330 Cham, Switzerland

Preface

This book is an attempt, from an essentially hydrogeological point of view, to bring together both traditional and more recent concepts relating to karst landscapes. If we are not to overlook the fundamental scientific aspects, this is the only most practicable approach.

Karst material has always aroused a great deal of interest due to its unique character and its huge economic importance, and also because the terrain that it creates often permits us to embark on underground exploration. This has been a major attraction over the years for all who love hidden things and is why the early scientific contributions had an exploratory component of the unknown, the underground world. Classical works are Martel's *Les abîmes*, Llopis' posthumous book and numerous journals around the world that disseminated the more or less sporting results of the exploits of many hardworking speleologists. Journals such as *Lapiaz* and *Spelunka* are among many that supported thousands of topographic surveys and records. The International Speleological Union (UIS) brought together illustrious sportsmen and women who loved the subterranean world, as well as scientists passionate about the land's depths, to engage in activities that are not without risk and that have left many behind in the caverns and passages that they so wanted to discover.

Scientists have entered fully into this exciting domain, investigating morpho-dynamic, morphogenetic and hydrogeological aspects in not only theoretical but also highly applied research. Karst water has a clear economic interest and is often the best—or only—source of urban water supply. The International Association of Hydrogeologists has been active in this regard, founding the Commission for Karst Hydrogeology. This has published numerous core reference works and organized milestone events in the advancement of this unique field of science. The Dubrovnik Colloquium (1966) initiated the series that continued in meetings in Antalya (Turkey), Besançon and Neuchâtel, alternating for two decades, also in Nerja and Malaga. More details can be found at http://www.iah.org/karst/, http://www.speleogenesis.info/, http://www.karstportal.org/, http://karstwaters.org/, http://nckri.org/, http://www.karst.edu.cn, http://www.irck.edu.cn and http://www.sedeck.org/.

Much of this information, together with research by numerous centres (Laboratoire Souterrain de Moulis of the CNRS, University of Neuchâtel, and Granada, Besançon, Montpellier and Malaga Universities), has formed the basis of this work, which aims to serve as a route into the exciting world of karst hydrogeology. It aims to provide a foundation that I hope will help the interested graduate to come to know better its characteristics, the parameters of interest and their quantification and the means for exploration, acting as a solid base of knowledge on capturing and exploiting this water resource. This volume also details the physical and chemical characteristics of karst water and specifies how it may be protected from contamination.

This book is the result of many years of dedication to the study of the hydrogeology of karst, at first in a personal way and later with the decisive support of numerous collaborators, some of them already university professors and other professionals. My work has benefited from a wide range of funding, both public and competitive—national, regional and European—as well as research contracts of many types.

Degree Theses

Benavente, J. 1978. *Hydrogeological research in the Sierra de Jaen.*
Casares, J. 1978. *Hydrogeological investigations in the karstic massifs of Parapanda and Hacho de Loja (province of Granada).*
Fernández, R. 1980. *Hydrogeological investigations to the North of Ronda (Málaga).*
Moreno, I. 1981. *Contribution to the hydrogeological knowledge of the Sierras de Maria and Maimon (province of Almería).*
Obartí, F.J. 1986. *Systems analysis applied to karst hydrogeology.*
Calaforra, J.M. 1987. *Hydrogeology of the karstified gypsum of Sorbas (province of Almería).*
Molina, L. 1989. *Contribution to the hydrogeochemical knowledge of the eastern sector of Campo de Dalías (Almería).*
López-Chicano, M. 1989. *Geometry and structure of a perimediterranean karstic aquifer: Sierra Gorda (Granada and Malaga).*

Doctoral Theses

Padilla, A. 1990. *Application of mathematical models to the study of karstic aquifers.*
Navarrete, F. 1992. *Contribution to the hydrogeochemical knowledge of Campo de Dalías.*
López-Chicano, M. 1992. *Contribution to the knowledge of the karstic hydrogeological system of Sierra Gorda and its surroundings (Granada and Malaga).*
Calaforra, J.M. 1996. *Contribution to the knowledge of gypsum karstology.*
Martín-Rosales, W. 1997. *Effects of the check-dams on the southern edge of the Sierra de Gádor (Almería).*
Andreu, J.M. 1997. *Contribution of overexploitation to the knowledge of the karstic aquifers of Crevillente, Cid and Cabeço d'Or (province of Alicante).*
Vallejos, A. 1997. *Hydrogeochemical characterization of the recharge of the Campo de Dalías aquifers from the Sierra de Gádor (Almería).*
Molina, L. 1998. *Hydrochemistry and marine intrusion in Campo de Dalías (Almería).*
El Morabiti, K. 2000. *Contribution to the geological, hydrochemical and isotope knowledge of the thermal waters of northern Morocco.* Univ. Abdelmalek Essaadi, Tetouan.

Contreras, S. 2006. *Spatial distribution of the annual water budget in semi-arid mountain regions. Application in Sierra de Gádor (Almería).*
Daniele, D. 2007. *Application of geographic information systems to the study of complex aquifers. Case of Campo de Dalías.*

Finally, I list the main research projects with which I have been involved, as well as the most relevant contracts.

Research Projects and Contracts

Mathematical models applied to the analysis of karst aquifers. CAICYT, 1983–1987.
Overexploitation in karstic aquifers. DGICYT. 1988–1992.
Hydrogeological aspects of groundwater protection in karst areas. CICYT. 1992–1995.
Characterization of contaminating processes in karstic aquifers. CICYT. 1995–1998.
Hydrogeological characterization of karst aquifers in semi-arid regions. The case of the Turón–Sierra de Gádor macrosystem. PO6-RNM-01696 Consejería de Innovación, Junta Andalucía, 2007–2010.
Action COST–65. *Hydrogeological aspects of the protection of groundwaters in karstic areas,* 1991–1995. 16 European countries.
Analysis and modelling of the elements of karst springs with a view to their characterization and forecasting of temporal evolution. Hispano–Bulgarian Project (CSIC—Bulgarian Academy of Sciences). 1991.
Comparative analysis of karst aquifer structures. Hispano–Bulgarian Project (CSIC—Bulgarian Academy of Sciences). 1992.
Mathematical simulation of the karstic coastal aquifers of Pinar del Río, Havana and Matanzas (Cuba). Institute for Ibero-American Cooperation (ICI).
CNIC Ministry of Higher Education, Cuba. 1994. *Mathematical simulation of the coastal aquifer of Zapata, province of Matanzas, Cuba.* ICI. CNIC. 1995.
Les Rencontres Méditerranéennes du Karst, EU, DG XI/A/2 France, Portugal and Spain. 1995.
Ecological problems of karst waters caused by overexploitation and contamination (on the example of North-East Bulgaria). CIPACT930139, UE, COPERNICUS. National Institute of Meteorology and Hydrology, School of Mines, and Hydrocomp Ltd, Sofia, Bulgaria. 1994–1997.
Groundwater karst systems: Conceptual modelling and evaluation of their vulnerability. EST. CLG975809 NATO. National Institute of Meteorology and Hydrology in Sofia, Bulgaria. 1999–2001.
Monitoring and densification of the retention check dams on the southern edge of the Sierra de Gádor and analysis of their influence on the environment. Contract IARA—University of Granada. 1990–1993.
Hydrogeological study of the Fuente del Rey (Manantial de la Salud) and its surroundings (Priego de Córdoba). University of Granada–Priego de Córdoba Town Hall. 1992.
Hydrogeochemical study of the aquifer systems of the South of the Sierra de Gádor–Campo de Dalías. Contract Cajamar—University of Almería. 2001–2002.
Evaluation of recharge and proposals to increase infiltration in the aquifers of the South of the Sierra de Gádor–Campo de Dalías. Contract Cajamar—University of Almería. 2001–2002.
The hydrogeological problem of the Valle de Abdalajís Tunnel and its surroundings. Contract U.T. E. Ayegeo Abdalajís—University of Almería. 2005–2006.
Hydrogeological advisory services for underground works on the High-speed South line. Contract ADIF—Universidad Almería. 2009–2012.
41986, n° 6/52, *Research on the karstic hydrogeology of carbonate massifs.* C. Romariz (U. Lisboa) and A. Pulido-Bosch (U. Granada).

1991, *Comparison of hydrogeological and hydrogeochemical aspects in karstic aquifers linked to gypsum*. P. Forti (U. Bologna) and A. Pulido-Bosch (U. Granada).

1991, *Quantitative approach to karst hydrogeology. Comparative study of some Pyrenean and Betic karsts*. G. de Marsily (U. Pierre and Marie Curie, Paris) and A. Pulido-Bosch (U. Granada).

1991, *Comparative analysis of the structures of karstic aquifers*. C. Drogue (USTL, Montpellier) and A. Pulido-Bosch (U. Granada).

1991, Hispano–Bulgarian bilateral project *Analysis and modelling of the elements of karst springs with a view to their characterization and forecasting of temporal evolution*. D. Dimitrov (Institute of Meteorology and Hydrology, Bulgarian Academy of Sciences, Sofia) and A. Pulido-Bosch (IAGM, CSIC).

1992, Spanish–Bulgarian bilateral project *Comparative analysis of karst aquifer structures* (CSIC —Bulgarian Academy of Sciences).

My good speleologist friends have helped me so much at various points in my professional career: among others, Juan de Dios Pérez Villanueva, doctor in Geography and firefighter; and Toni Fornes, who provided me with numerous photographs of the Vallada area for this book, to mention just two especially representative of this generous group of individuals who love karst so much. The final phase of the book was written in the Department of Geodynamics of the University of Granada during a long personal stay that had the support of the members of the department, especially Profs. Calvache and Azañón. I wish to express my sincere gratitude to them all for their remarkable efforts and the sincere friendship that they have always extended to me.

This work would not have been possible without the continued help of Paule Leboeuf Gaborieau over many years and his enormous effort in adapting all the figures that are included in this book. It is clear that this is a joint work. Thank you so very much.

Finally, I would like to acknowledge the contribution of Dr. Alexis Vizcaino (Springer Nature) and Alison Williamson (Burgess Pre-Publishing) editing this textbook. The English edition is based on *Principios de Hidrogeología kárstica*, published by Editorial Universidad de Almería in Spanish in 2015.

Granada, Spain   Antonio Pulido-Bosch
March 2020

Contents

Karst and Pseudokarst Materials

1

1.1 Glossary

Karst has two meanings: the first is synonymous with **karst region**, *one made up of carbonate, compact and soluble rocks that display characteristic surface and underground forms*; the other, by extension, refers to any effect of karstification on karstifiable rock.

Pseudokarst alludes to a region that presents forms analogous to those of karst in rock that is only slightly or not at all karstifiable (subject to karstification).

Karst phenomena refer to all karst forms and the processes that determine them; the latter comprise **karstification**.

Karst material is material from **karst**; in the wider sense, it is used to refer also to **pseudokarst** and **thermokarst** material.

Microscopic voids between minerals create **intercrystalline porosity** of 0.1 to 1% of the rock's total porosity. The voids between cemented grains are **interstitial porosity**. Both can be referred to as **matrix porosity**, as opposed to **conduit porosity**.

Macroscopic porosity refers to large karstified fractures, conduits, channels and caves; a specific case is **cavernous porosity**, when large holes of karst origin predominate.

Dolomitization is mole-by-mole replacement of the calcium in a limestone by magnesium, which results in a volume reduction of about 13% as the dolomite is denser than calcite (2.866 against 2.718 g/cm^3). This leads to an increase in the porosity of the resulting dolomitic rock. This type of **dolomitization** porosity is **intercrystalline.**

Interconnected porosity (Po) is the total volume of interconnected pores, effectively equivalent to total porosity.

Relative drainability equates to $S_o = S/Po$, the quotient of a rock's storage coefficient (S) and its coefficient of open porosity (P_0).

© Springer Nature Switzerland AG 2021
A. Pulido-Bosch, *Principles of Karst Hydrogeology*, Springer Textbooks in Earth Sciences, Geography and Environment,
https://doi.org/10.1007/978-3-030-55370-8_1

1.2   General Aspects

The term 'karst' derives from the region between Trieste and Ljubljana (Laibach), as it formerly belonged to the Austro-Hungarian Empire (Karst, until 1918), then Italy (Carso, until 1945), Yugoslavia and present-day Slovenia (Kras). This area has similar characteristics all along the eastern Adriatic (Istria, Croatia and Dalmatia) and is comprised of limestone with peculiar morphological and hydrological features. From a hydrological point of view, karst areas have an almost total absence of surface drainage, and they feature both endorheic basins and considerable groundwater circulation.

Classically, the term **karst** has two meanings. One is synonymous with karst region, a *region constituted of carbonate, compact and soluble rock with characteristic surface and underground forms*; the other, by extension, refers to the effect of karstification on a karstifiable rock. The term may also be applied to any region constituted of a soluble rock (gypsum and salt), for which some authors reserve the term *pseudokarst*, defined as a *region that presents forms analogous to those of karst in a rock that is only slightly or not at all karstifiable* (or the effects of karstification in such material). The concept of a *karst phenomenon* applies to both karstic forms and the processes that make them; the latter comprise *karstification*.

The importance of carbonate rock is clear from the fact that it represents around 5% in volume of the Earth's lithosphere. The percentage of carbonate rock of all sedimentary rock is approximately 15%. Carbonates predominate among relatively recent formations, because these are mainly organic sediments. Thus, we can estimate that approximately, 12% of the continental surface is comprised of carbonate rock. About 25% of the world's population is supplied by karst water [1–3]. In Europe, carbonate outcrops make up over 3 Mkm^2 (35%) of the land. Figure 1.1 shows the main expanses on both sides of the Mediterranean.

Spain's limestone regions cover about 100,000 km^2 (Fig. 1.2): 17,000 km^2 in the Cantabrian Cordillera, the Basque Country and the Pyrenees; 48,000 km^2 in the Iberian Cordillera; 7500 km^2 in the Catalan Cordillera; and 30,500 km^2 in the Betic Cordillera [4, 5]. It is estimated that the average annual hydrological recharge is 20,000 hm^3, while the reserves can exceed 200,000 hm^3, hence their enormous economic and ecological interest.

1.3   Classification and Composition of Karst Materials

A classic classification of *karst material* in *s. l.* is as follows (Table 1.1).

From both practical and economic points of view, karst material in its narrow sense is of most interest. Hypersoluble karst material will also be described—specifically gypsum, since the remaining substances are studied by other fields. The limestones and dolomites that constitute carbonate rock are karst material par excellence. The marls, a mixture of carbonate and clay, are of little interest as aquifers. The four aspects of a carbonate rock usually considered are: its chief

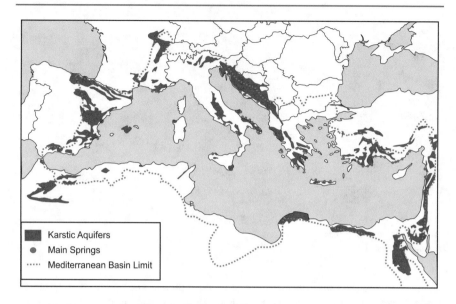

Fig. 1.1 Main outcrops of carbonate materials in the Mediterranean Basin (Adapted from [6])

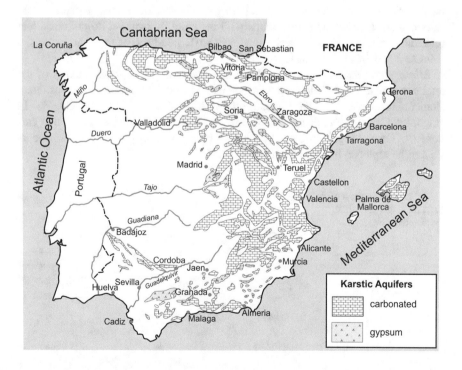

Fig. 1.2 Main outcrops of karst materials in Spain (Adapted from [7])

Table 1.1 Division of karst materials

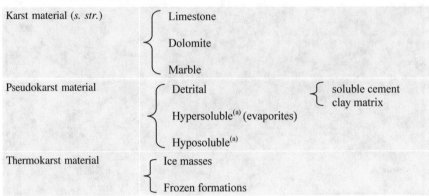

Karst material (*s. str.*)	Limestone
	Dolomite
	Marble
Pseudokarst material	Detrital — soluble cement / clay matrix
	Hypersoluble[a] (evaporites)
	Hyposoluble[a]
Thermokarst material	Ice masses
	Frozen formations

[a]With regard to limestone
Adapted from [8]

element; its minority and trace elements; its fluid inclusions; and its non-carbonate components. The main minerals are calcite, dolomite and gypsum, but there are many others (Table 1.2).

As the radius of Ca^{2+} is 0.99 Å and Mg^{2+} is 0.66 Å, the substitution of one for the other leads to a considerable change in a rock's volume. Further carbonates are able to make isomorphic substitutions with the chief mineral yet do not constitute rock themselves, such as ankerite ($CaFe(CO_3)_2$), siderite ($FeCO_3$), rhodochrosite ($MnCO_3$) and witherite ($BaCO_3$).

Minor elements and traces are those that make up just a small fraction of the total constituents. The following elements can replace a mineral's main constituent element. In calcite, Ca can be replaced by: Mg, Mn, Fe^{2+}, Sr, Ba, Co, Zn. In aragonite, Ca can be replaced by: Sr, Pb, Ba, Mg, Mn. Dolomite is a double carbonate of Ca and Mg: its Mg can be replaced by Fe, Mn, Pb, Co, Ba, Zn, Ca; and its Ca can be replaced by: Mn, Fe, Pb and Al.

Table 1.2 Main elements in karst rock, showing their characteristics

Mineral	Density	Composition	System
Calcite	2.71	$CaCO_3$	Trigonal
Aragonite	2.94	$CaCO_3$	Rhombic
Dolomite	2.87	$MgCa(CO_3)_2$	Hexagonal
Magnesite	3.06	$MgCO_3$	Hexagonal
Gypsum	2.32	$CaSO_4.2H_2O$	Monoclinic
Anhydrite	2.95	$CaSO_4$	Rhombic
Polyhalite	2.78	$K_2Ca_2Mg(SO_4)_4.2H_2O$	Triclinic
Halite	2.16	$NaCl$	Cubic
Silvinite	1.99	KCl	Cubic
Carnalite	1.6	$KCl.MgCl_2.6H_2O$	Rhombic

Inclusions, on the other hand, can be detected under the electron microscope in the form of tiny bubbles or droplets. When abundant, these give a spongy appearance to certain calcites. The droplets may contain Na^+, K^+, Cl^-, together with Ca^{2+}, Mg^{2+} and $SO_4^=$, besides gases such as CO_2 and CH_4. Under certain conditions, their study yields valuable information on the genesis of the minerals and the environmental conditions, hence is of interest when prospecting for mineral deposits and undertaking palaeoclimatic reconstruction.

Of the *non-carbonate components*, the clay fraction is the most plentiful and significant impurity; silica is also abundant, both of detrital origin and chemical precipitation (nodules or strata). Other non-carbonate minerals that may be present are: fluorite, celestine, zeolite, göetite, barite, phosphate, pyrolusite, gypsum, stroncyanite, feldspar, mica, quartz, rutile, glauconite—also chlorite—tourmaline and pyrite-marcasite. This explains the presence of these ions in solution, not necessarily due to contamination.

The minerals listed above are the main ones seen in karstification and likely to be found in karst caves, but the actual list is much longer. Hill and Forti [9] have edited an interesting collection of slides of the very wide range minerals found in caverns and speleothems.

1.4 Porosity and Permeability in Carbonate Rock

1.4.1 General Aspects

The high porosity of a carbonate mud is reduced to a tiny fraction of its original by diagenesis, in consequence of the processes of compaction, cementation and recrystallization, from 5 to 15% of total porosity in the most favourable cases. Its permeability, according to investigations in hydrocarbon prospecting, will also be very low; the highest values are in calcarenite and calcirrudite with little cementation and in highly recrystallized dolomite, at between 10^{-3} and 10^{-7} cm/s. Therefore, these materials' permeability ranges from low (in an aquifer) to impervious (in an aquifuge).

Several secondary processes significantly increase the effective porosity of a material. These include secondary dolomitization, provided it takes place when the sediment has been consolidated; otherwise, it has very little effect. In this case, the change from calcite to dolomite creates 13% of the rock's total porosity. Porosity is reduced by recrystallization and is enhanced by selective leaching processes.

At least four types of porosities are usually recognized ([10], Table 1.3), depending on the nature of the voids likely to store gravity water. *Microscopic voids* between the minerals generate an *intercrystalline porosity* of up to 0.1–1% of the rock's total porosity. Voids between cemented grains create what is known as *interstitial porosity*. In this way, some types of limestone (oolitic, for example) have an interstitial porosity similar to that of a sandstone. The *porosity of microfissures* corresponds to the voids created mostly by microcracks, diaclases, stratification

Table 1.3 Classification of voids and types of porosity

Scale	Void		Porosity type
Microscopic	Pores or interstices	Intercrystalline	Intercrystalline
		Intergranular	Interstitial
	Microfissures	Stratification joints	Microfissures
Macroscopic		Conduits	Conduits

joints and schistosities. *Macroscopic porosity* refers to large karstified conduits, channels and caves, where fractures have been preferentially exploited by karstification; a specific case is *cavernous porosity*, where holes of karst origin predominate [11].

The first three porosities described above can be generically grouped under the heading of *matrix* porosity, as opposed to *conduit* porosity and that derived from *fractures*. Classical studies of petroleum geology conclude that the greatest total porosity is seen in highly recrystallized calcilutites, followed by pisolitic uncemented limestone, bioclastic limestone with a dolomitic microgranular matrix (15%), oolitic microgranular dolomitic matrix limestone (12%) and microcrystalline dolomite (11%). Values below 2% correspond to micritic limestone and all well-cemented variants.

1.4.2   Porosity and Permeability in Dolomites

1.4.2.1   Types of Porosity in Dolomite

Dolomitization Porosity

Dolomite, in general, is a porous and permeable rock and hence is both a good aquifer and an oil storage rock. Most dolomites are secondary rocks, formed from pre-existing limestone by replacement of part of its Ca by Mg. In the natural environment, *direct precipitation* of dolomite, to create *primary dolomite,* is almost non-existent. The few known examples, of limited extent, are in warm, shallow lagoons that are supersaturated in Mg^{2+}, supratidal plains along extremely arid coasts, known as *sabkhas*, and hypersaline lakes. Replacement dolomite, on the other hand, is generated throughout diagenesis from its earliest to its latest stages and can either affect the entire mass of sediment and/or original limestone rock or take place selectively. In the absence of any extra contribution of carbonate ions and calcium ions to the starting content of the limestone, the mole-by-mole replacement causes a reduction in volume of about 13%, as dolomite is denser than calcite (2.866 compared to 2.718 g/cm^3). This results in an increase in the porosity of the resulting dolomitic rock. This type of porosity, known as **dolomitization porosity**, is *intercrystalline* (Fig. 1.3), and the rock consists of a network of partially interpenetrated and relatively well-formed dolomite rhombuses in a structure known as 'sacaroid dolomite' or 'sugar dolomite'.

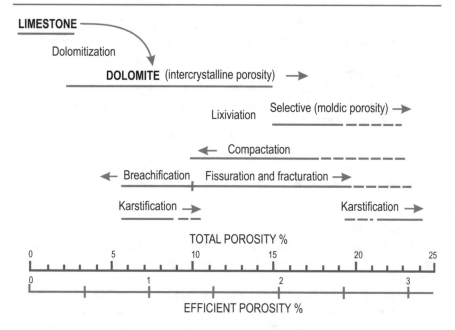

Fig. 1.3 Conceptual scheme of the development of dolomite's porosity as a function of a series of processes (Adapted from [12])

Also of great interest is the porosity that arises during dolomitization by *selective lixiviation* of some textural elements of the original limestone, due to the fact that the dolomitizing fluid is initially undersaturated or partially subsaturated with bicarbonate ions. This type of porosity is *moldic* and very similar in magnitude to its effective porosity. There are good examples in the Betic Cordillera [13] in the ochre dolomite of the Prebetic Upper Cretaceous and the dolomitized reefs of the Mediterranean Messinian. In the first, the presence of dissolution voids formed by preferential lixiviation of some types of fossil fragments (bioclasts) during dolomitization confers a high porosity on the rock, from 5 to 20% of its total volume.

In the example of the Messinian reefs, the porosity caused by the lixiviation of textural elements of the aragonite's original composition, which takes place throughout dolomitization, adds to the reef's initial porosity, for example that inherent in the bio-building framework and the coral breccia of its upper slopes. When the two are combined, the porosity can be greater than 20% of the rock's overall volume. In general, the open spaces in the original limestone are usually preserved in all dolomitization, resulting in high porosity values when added to that created by dolomitization processes.

Frequently, the porosity caused by dolomitization, especially intercrystalline porosity, is significantly reduced or even cancelled out by subsequent **compaction**. This is especially the case when there has been incomplete dolomitization, as the lattice of dolomite crystals does not then offer enough rigidity to counteract the compaction. Late **cementations**, often of sparitic calcite, may also reduce porosity.

Fracture Porosity

Dolomites are fragile rocks and often appear highly fractured in outcrops, crossed by a multitude of small fissures. The extent of the increase in porosity from this fracturing is easily quantifiable. However, if it is very intense, the rock breaches (Photos 1.1 and 1.2). Such is the case of the *Alpujarrides* dolomite, widespread in the southern provinces of Malaga, Granada and Almería, which has very low permeability values. It seems that once a threshold has been reached, the dolomite breaks and its porosity and permeability do not so much increase as diminish considerably.

Karstification Porosity

Dolomite terrain can be a karst aquifer. The karstification of dolomite, which exploits the previous openings, as in limestone, has the effect of considerably widening any cracks and effectively increasing the rock's porosity. Dolomitic rocks are soluble. Although dolomite is more soluble than calcite and Mg is more soluble than Ca, dolomitic outcrops are leached less than limestone, because dolomite karst water soon becomes saturated and does not precipitate dolomite as speleothems. In limestone, on the other hand, although saturation point is reached temporarily in the

Photo 1.1 Crushed dolomites or *kakiritas* of the western edge of Sierra Nevada, used as building material (Photo A. Pulido)

Photo 1.2 Detail of another brecciated dolomite (Creu formation, Valencia; Photo A. Pulido)

karst water, it precipitates calcite frequently as speleothems in caves and other cavities, thus returning to a subsaturated state that can again dissolve the rock.

When karstification is selective and acts on partially dolomitized limestones or partially calcified—de-dolomitized—dolomite, *carniolar structures* develop. In the Betic Cordillera, the most important carniolar outcrops are at the base of the Lower Liassic dolomite, and they constitute very interesting aquifers. These particular carnioles correspond to ancient dolomitic rocks that have been partially de-dolomitized by waters rich in calcium sulphate from the leaching of evaporites (gypsum) from the underlying Keuper. Subsequent erosion has acted differentially, releasing the dolomitic portions that were more soluble and mechanically less resistant, giving rise to its characteristic vacuolar structure.

1.4.2.2   Permeability

The permeability of dolomite is very variable, as it is a function of many factors, but is generally considered to be far more constant than in limestone terrain. Extensive data have been obtained from oil research, which regards dolomite as a good rock store. The highest permeability values are observed in dolomites with a high porosity—whether moldic or sacaroid, and fissured—that are unbrecciated and also strongly karstified (Fig. 1.3). The highest values, at over 1000 millidarcies, are seen in highly crystalline dolomite.

1.5   Porosity and Permeability in Matrix

The investigation of the hydrogeological properties of carbonate material can help to establish a conceptual model of such aquifers. In any conceptual model, the matrix plays an important role, especially with regard to the time taken to empty an aquifer, the extent of its residual saturation and the time for it to replenish itself, in the case of an overexploited aquifer [14, 15]. Probably, the main importance of matrix is to the identification of any pollution of the aquifer and subsequent correction [16].

Conceptual models of carbonate aquifers [17–19] often ignore the role of matrix or consider it to be minor, laying more emphasis on fractures and subsequent karstification. However, hydrodynamic and/or hydrogeochemical anomalies are best understood by taking into account the matrix, and the interpretation of a tracer test [20, 21] is quite different if its porosity is considered. For these reasons, studies to evaluate the hydrogeological properties of carbonate matrix are of considerable interest [22].

To illustrate such research, after examining the analysis of fractures, we will comment on two studies on the matrix of carbonate rocks: one on the Devonian of the Olkusz region [22] and one on the Betic Cordillera [23, 24], also the sector south of Ronda, Malaga, the Sorbas gypsum and the Sierra Gorda (see Boxes 1, 2, 3, 4 and 5).

1.6   Fracture Analysis

1.6.1   Basic Concepts

Discontinuities are the points of access and initial circulation of groundwater, in which fracturing and fissures play a key role (Photo 1.3). For this reason, it is interesting to carry out a detailed study of fracturing.

The methodology comprises at least two aspects:

- Location of fractures on aerial photo. Depending on the scale (flight height), the information comes from large alignments, medium alignments or small to very small fractures. While there is a marked subjective component, the method yields valid statistical results. The subsequent study technique can be manual, using 'optical filtering' (optical bench), or digital, which is most common currently. It is possible to work either with the frequencies of the trends, with lengths being added on the basis of their direction, or with the separation between fractures, essentially. Fracturing can be shown to be a random variable with a structural component.
- To achieve the correct interpretation, the work on aerial photo has to be complemented by measures on the ground. Aerial photo indicate only the line of a structure's intersection with the topography, with no clues to its dip, so

Photo 1.3 Detail of dense fracturing in the Creu formation in Barranco del Infierno, beside the River Serpis (Valencia. Photo A. Pulido)

quarries and either natural or artificial *galleries* are an invaluable source of this information.

In areas of little tectonic complexity, the study of associated minor structures— stylolites, stress cracks and fault drag—alongside photogeological study can determine the angle of the stress ellipsoid. Studies have shown that fracture orientation is uniform across wide areas and is maintained at depth, and that, the frequency of fractures of the same alignment depends upon lithological heterogeneity—the presence of alternating marl strata, for example—and on the thickness of the strata: the less the thickness, the greater the fracture density.

On the other hand, water circulation is not directly related to the density of the fractures so much as to the density of fractures that have had one or more episodes of tension (opening) in their history, provided that clogging and compaction are not involved. The development of fissures guarantees a physical continuity in the circulation environment, permitting the presence of *aquifers* in a carbonate terrain. Karstification processes are superimposed on this underlying framework, further developing a set of *open* fractures to the detriment of others and generating block structures of very varied size as a function of the fracturing. This can be at the scale of tens or hundreds of metres across, or kilometres wide.

Wittke and Louis demonstrated that $k = \frac{gd^3}{12v}$ where g = gravity acceleration; d = fissure opening; and v = kinematic viscosity. This aspect is fundamental, since it is not so much the fractures' number, orientation and frequency as their separation that is the key, to the extent of the power of three. Furthermore, separation determines not only the *k-value* but also the type of flow, whether laminar or turbulent, parallel or otherwise, and the water circulation. For example [18], given that suction relies on the narrowness of fissures, controlled by surface tension, the height at which suction operates varies from 300 cm for fissures 5 microns across, 150 cm for cracks 10 microns wide and only 0.15 cm for openings of 1 cm, as these last permit turbulent flow.

Kiraly [25] proposes an expression to estimate the value of permeability k as a function of the frequency of each family (f_i) and the separation of fractures (d), taking into account the matrix identity I $k = \frac{g}{12v}\sum f_i d_i^3 [I - n_i * n_i]$. For his part, Müller proposes the expression $me = \sum f_i d_i$ to estimate the effective porosity (me) according to the frequency of each family (f_i) and fractures' separation (d_i). Rats [26] establishes a relationship between interfractural separation (d) and the thickness of strata (e), arriving at the logarithmic expression $log d = a + b\, log e$, with $a = -0.64$; $b = 0.41$.

To illustrate the above, we will apply fracture analysis to the karst aquifers of the Serranía de Ronda [27] (Box 3) and the Sierra Gorda aquifer [28, 29] (Box 4), with the simple intermediate example of the Sorbas gypsum [30].

1.6.2   Box 1: Olkusz Region (Poland)

Hydrogeological investigations were carried out on samples from four boreholes in the Klucze region (Fig. 1.4), about 8 km north of Olkusz. Samples from the different boreholes were taken at depths of between 180 and 500 m. The samples are mainly limestone and dolomite (Muschelkalk facies) and Jurassic, with only one sample of marl. The diameters of the sample cores were between 46 and 47 mm, and their heights between 47 and 56 mm.

Interconnected porosity (Po), the total volume of interconnected pores (equivalent to total porosity) and the storage coefficient (S; *specific yield*) (effective porosity, actually) were measured in 127 samples of limestone, 37 of dolomite and one of marl. Permeability (k) was measured in 126 samples of limestone and 37 of dolomite. The k-values were calculated in the samples using air as the fluid, so they were later recalculated for water at a temperature of 10 °C.

A vacuum chamber was used to measure the open porosity by extracting all the air from the sample and filling the empty space with water while carrying out a series of measurements of its weight. This evaluates the volume of interconnected pores. Prior to going in the vacuum chamber, the samples are dried for at least 24 h in an oven between 105 and 110 °C. The following formula [31, 32] is used to calculate the open porosity coefficient (P_0):

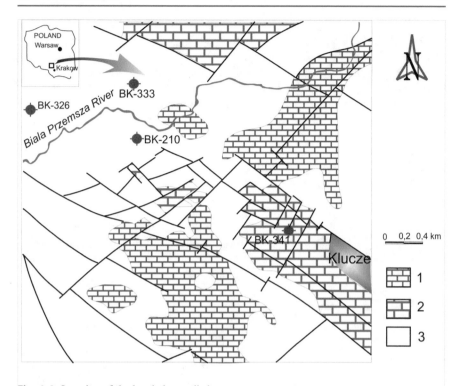

Fig. 1.4 Location of the boreholes studied

$$P_0 = \frac{G_n - G_s}{G_n - G_{nw}} \tag{1.1}$$

where G_n is the weight of the water-saturated sample, G_s the weight of the sample dried at 105–110 °C, G_{nw} is the weight of the water-saturated sample, weighed into the water—using Archimedes principle.

The methodology used to calculate the storage coefficient (S) uses a centrifuge to extract water to accelerate the task, as the natural release of gravity water is a very slow process. The suction pressure exerted on the sample by centrifugal force releases a certain part of the total volume of water contained in a sample (known as gravity water), which is calculated by the following formula:

$$H = \frac{\left(\frac{2\pi n}{60}\right)^2 r h}{g} \tag{1.2}$$

where H is the water suction pressure of the matrix, expressed in metres of height of the water column; n is the number of revolutions per minute; r is the centrifuge radius (distance in m from the axis of the centrifuge to the centre of gravity of the sample); h is the length of the sample in m; and g is the acceleration of gravity (9.81 m/s^2). The challenge is to set the appropriate suction pressure to simulate natural conditions; in this method, it is the equivalent of a 10 m high water column. The rock's storage coefficient is based on the quantity extracted. Determining the variables ($H = 98$ kPa and length of each sample) derives the number of revolutions to be applied, and the volume of water obtained from the centrifuge is then used to calculate the storage coefficient (S):

$$S = \frac{V_w}{V_r} \tag{1.3}$$

where V_w is the volume of water released for a suction pressure equivalent to a water column 10 m high (cm^3) and V_r is the volume of rock (cm^3). The simulated water extraction pressure in the centrifuge for small samples is equivalent to the maximum water extraction pressure in nature by the action of gravity on a stratum of thickness h.

Centrifugal acceleration (a) is expressed as

$$a = \omega.R^2 \tag{1.4}$$

where a is centrifugal acceleration and ω angular velocity. Working with Eq. 1.2, we arrive at:

$$\frac{H}{h} = \frac{a}{g} \tag{1.5}$$

and also:

$$a = \left(\frac{2\pi n}{60}\right)^2 r \tag{1.6}$$

so that n (no. of revolutions per minute) can be calculated for each sample. According to Prill et al. [32], the relationship between the time for the percolation of gravity water in nature (T_n) and the time spent in the centrifuge (t) can be expressed by:

$$\left(\frac{T_n}{t}\right) = \left(\frac{a}{g}\right)^2 \tag{1.7}$$

All samples were centrifuged for 30 min (T) which, depending on their length, is equivalent to a percolation time under natural conditions (T_n) of between 660 and 940 days (from 2 to 2.5 years). To verify that 30 min is sufficient to allow the removal of the gravity water from the sample, a group of 25 samples were tested; the rest did not release anything in the centrifuge.

As a result, 21 of the sample's results lay between the fastest and slowest graphs, while the remaining four gave virtually no water. Thus, the process takes a maximum of 11 min.

A new parameter known as relative drainability (S_0) can be defined:

$$S_0 = S/P_0 \qquad (1.8)$$

as the quotient of the storage coefficient (S) and the open porosity coefficient (P_0). This gives an idea of the diameter of the pores of the matrix, as well as their nature (fissured or small capillary, among others).

For permeability analysis, samples are dried at 105–110 °C and then put into the air permeameter. The expression that allows the calculation of Darcy's permeability coefficient (K_g), expressed in darcys, is as follows:

$$K_g = \frac{2 \cdot Q_0 \cdot P_0 \cdot L \cdot \eta}{F \cdot \left(p_1^2 \cdot p_2^2\right)} \qquad (1.9)$$

where Q_0 is the gas flow (cm³/sg); p_0 is the atmospheric pressure (atm); L is the length of the sample (cm); η is the dynamic viscosity coefficient of the gas; F is the sample section (cm²); p_1 is the gas pressure before passing through the sample (atm); and p_2 is the gas pressure after passing through the sample (atm). The coefficients obtained are then recalculated for water at 10 °C temperature (K_{10}), according to the equation:

$$K_{10} = K_g \frac{\gamma}{\eta} \qquad (1.10)$$

where γ is the specific weight of the water. After considering this, we arrive at:

$$K_{10} = 7.66 \times 10^{-6} \, K_g. \qquad (1.11)$$

However, the permeability calculated using this formula does not correspond to that which would be obtained naturally and using water. The *Klinkenberg correction coefficient*, which depends on many factors and is inherent to each type of rock, must therefore be employed. Consequently, we know that Kg and the recalculated for water (K_{10}) are in fact lower, especially in samples with low permeability.

The investigated samples were described in terms of microscopic features and by simple field methods. They are usually limestone and dolomite. Depending on their macroscopic structure and texture, the following types can be distinguished, with the number of samples of each in brackets: micritic limestone (39); fissured micritic limestone (58); conglomerate limestone (30); brecciated and fissured dolomite (32); recrystallized dolomite (5); and marl

(1). In fissured or fractured rocks, the filling is calcite. Many fractures, especially in the limestone, are filled with yellowish clay. Seven of the samples show the presence of stylolites [22].

The Devonian limestone of the Klucze region has relatively low open porosity. Its value varies between 0.00185 and 0.064. The mean is 0.0152, the standard deviation 0.014 and the coefficient of variation 0.92. The fissured, micritic and conglomerate limestone has the highest porosity, with the same average porosity. The lowest porosity is seen in the brecciated limestone.

The distribution of the values of open porosity is not homogeneous; there are two subgroups of a dissimilar distribution (Fig. 1.5). The first subgroup includes rocks that show a very low porosity, and the second those that have a slightly higher porosity. Open porosity in the dolomite is generally lower than in the limestone and varies between 0.00212 and 0.0259. The arithmetic mean is 0.0108, the standard deviation 0.00663 and the coefficient of variation 0.61. The highest porosity value is found in the brecciated dolomite with filled fissures. The crystallized dolomite shows a lower value, and the dolomitic breccia lower again. The distribution of open porosity in the dolomite is more homogeneous than in the limestone (Table 1.4). The only sample of a marl has an open porosity value of 0.181, which is higher than that of the limestone and dolomite.

The storage coefficient obtained for the Devonian limestone is very low: only seven of the 127 samples released traces of water. It varies between 0.00064 and 0.00163, and the mean is 0.000065. Although all 127 samples provided water, there are differences between the mean S values for the types of rocks described. The highest S value is shown by the micritic limestone with filled fissures and brecciated limestone, with an average value of

Fig. 1.5 Cumulative frequencies of porosity values

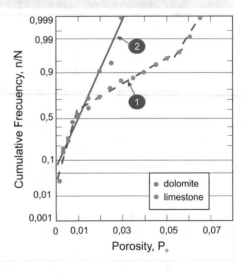

Table 1.4 Extreme and mean values of Pa, S and K in limestone and dolomite

Lithology	P_0 (%)		S (%)		k (m/s)	
Limestone	0.2–6.4	1.5	0–0.16	0.006	5.32×10^{-12}–4.71×10^{-10}	1.94×10^{-10}
Dolomite	0.2–2.6	1.1	0–0.66	0.13	1.3×10^{-11}–6.33×10^{-7}	2.7×10^{-10}

0.00011. The lowest mean value is found in the micritic limestone, at 0.0000164 (Table 1.4).

The dolomite's storage coefficient is somewhat higher than that of the limestone. Also, of the 37 samples of dolomite, 20 provided water, a much higher percentage than the limestone. The values ranged from 0.00032 to 0.0066, with an average of 0.00133. The distribution of this coefficient can be seen in Fig. 1.6. The sole sample of marl provided no gravity water, as expected.

In relation to S_0 (*relative drainability*), in the limestone, it ranges between 0.012 and 0.070, and in the dolomite between 0.025 and 0.80. The average value for all the limestones is 0.00208, and for all the dolomites 0.126. Rock type has no great influence on the value of the S_0 coefficient. The conglomerate limestone has the lowest value (0.00083), and the limestone with filled fissures and the brecciated limestone the highest (0.00527 and 0.00435, respectively). As for the dolomite, the recrystallized ones did not release water ($S_0 = 0$); in the brecciated dolomite, the mean value is 0.203, and in the filled fissures, it is 0.119.

The drainage dynamics is of great interest: it has been proven to be very fast. For the 25 limestone samples investigated, drainage ceases in most cases after 10 min in the centrifuge, equivalent to 240 days of natural drainage. Figure 1.6 shows this diagrammatically.

The permeability of the matrix of Devonian limestone and dolomite ranges widely from 5.32×10^{-12} to 4.71×10^{-8} m/s; the mean, before the

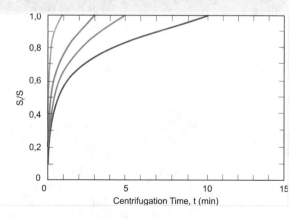

Fig. 1.6 Drainage dynamics of samples (Sandstone: total specific flow ratio—S- and instantaneous storage coefficient—Sr-)

Table 1.5 Means of hydrogeological parameters of the rock types studied: (1) Micritic limestone; (2) Micritic fissured limestone; micritic limestone with filled fissures and conglomerate limestone; (3) Brecciated limestone; (4) Brecciated dolomite and brecciated dolomite with filled fissures; (5) Recrystallized dolomite

Rock type	No. of samples	P_0	S	S_0	k (m/s)
1	39	0.0128	0.0000164	0.00161	1.41×10^{-10}
2	58	0.0182	0.000110	0.00487	0.71×10^{-10}
3	30	0.0127	0.0000420	0.000830	1.55×10^{-10}
4	32	0.0110	0.00154	0.145	3.29×10^{-10}
5	5	0.00986	0.00	0.00	9.31×10^{-11}

Klinkenberg correction, is 1.94×10^{-10} m/s. There are differences in permeability between the types of rocks described. The lowest value is seen in the micritic limestone (1.41×10^{-10} m/s), followed by the conglomerate limestones (1.55×10^{-10} m/s), the fractured limestones (2.69×10^{-10} m/s) and the brecciated limestones (2.13×10^{-10} m/s). The highest value is 5.65×10^{-10} m/s, displayed by the micritic fissured limestone.

The permeability in the dolomites is, in general, a little higher than in the limestones, ranging between 1.30×10^{-11} and 6.33×10^{-7} m/s, with an average of 2.77×10^{-10} m/s. The permeability in dolomite is a little higher than in the limestone, ranging between 1.30×10^{-11} and 6.33×10^{-7} m/s, with an average of 2.77×10^{-10} m/s. The brecciated types of dolomites display quite different values from the others, at 4.65×10^{-9} m/s. Crystallized dolomites and brecciated dolomites with filled fissures reach values of 9.31×10^{-11} (Table 1.5), and 9.86×10^{-11} m/s, respectively.

The distribution of the permeability values in limestone and dolomite is logarithmic but not homogeneous (Fig. 1.7). Two subgroups can be identified: the first has the limestones and dolomites with low permeability, probably micritic and with a relatively low number of fissures. The second has intensely fractured rocks with high permeability. It is difficult to estimate the influence of stylolites. There are only seven examples where these are present, and they have an average open porosity of 0.0190 and a permeability 4.01×10^{-10} m/s. There are no clear trends in relation to the variation in these parameters with the sample's depth. The most important differences are found when comparing samples from each borehole. In one borehole, high values of open porosity were noted in the depth range of 180–186 m, 305–315 m and 340–350 m, yet in another, they corresponded to 450–460 m and 495–500 m.

The basic physical property on which permeability and the storage coefficient depend is the geometry of the matrix. Figure 1.8a shows the relationship between open porosity and storage coefficient: there is no significant correlation between the two variables. Figure 1.8b represents the relationship between open porosity and permeability in limestone: there is a slight tendency for open porosity to increase with permeability, but there is no notable

Fig. 1.7 Cumulative
frequency values of hydraulic
conductivity

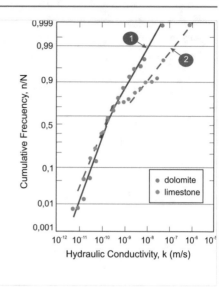

Fig. 1.8 a Storage versus
porosity coefficient (orange:
dolomite); b Permeability
versus porosity in limestone

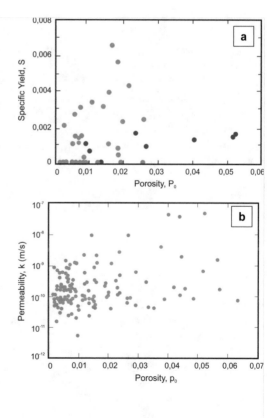

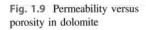

Fig. 1.9 Permeability versus porosity in dolomite

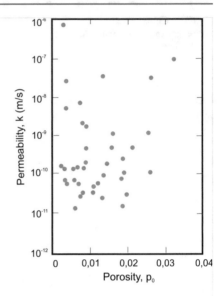

statistical relationship. Neither is there is any significant correlation between permeability and open porosity in dolomite (Fig. 1.9).

It can be concluded that the matrix of the Devonian carbonate rocks of the Klucze region does not have the capacity to store and transmit groundwater that is mobilized by gravity. However, these rocks do retain water; thus, the matrix plays a role in the migration of tracers and pollutants. The dolomite matrix, on the other hand, is slightly more conducive to storing and transmitting gravity water.

1.6.3 Box 2: Betic Cordillera

The analysed samples were taken from eight boreholes for earlier research relating to the construction of reservoirs in the eastern part of Andalusia (Fig. 1.10). Three are located in the internal zones of the Betic Cordillera, three in the external zones and the remaining two in the neogenic terrain of the intraorogenic basins, ranging in age from Palaeozoic to Cenozoic. The 181 samples were taken at depths of between 1 m and 258 m. Of these, 98 samples were of limestone or dolomite, 75 of calcarenite, six of marly limestone and two of marble. Their diameters varied between 46 and 47 mm, and their average length was 50 mm. Before studying the hydraulic characteristics of the borehole core samples, they were carefully washed to eliminate drilling residues.

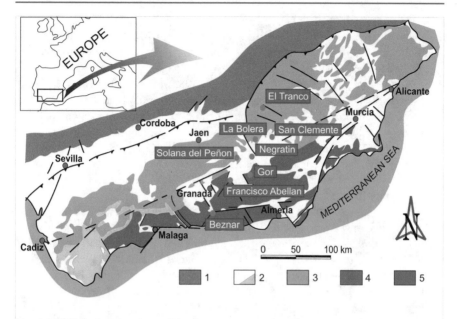

Fig. 1.10 Geological location within the Betic Cordillera of the reservoirs from which the samples were collected: (1) Foreland; (2) Neogene basins, volcanism; (3) External zone; (4) Flysch; (5) Internal zone (Adapted from [39])

Table 1.6 Basic statistics of properties samples from the Betic Cordillera

Statistics/parameter	P_0	S	S_0	k (m/s)
Average	0.05277	0.00579	0.0758	3.24E−7
Variance	0.002841	0.000138	0.016	1.58E−11
Standard dev.	0.0533	0.0118	0.126	3.98E−6
Minimum	0.004458	0.0	0.0	2.71E−12
Maximum	0.2091	0.0798	0.666	5.35E−5
Range	0.2046	0.0798	0.666	5.35E−5
Kurtosis	4.001	31.7	15.5	496
Variation coefficient (%)	101	203	167	1230

Table 1.6 summarizes the average statistics of the properties analysed in the 181 samples.

The *interconnected porosity* shows a wide variation. The greatest porosity was measured in the calcarenite and the minimum in the dolomite, but the minimum average value was in marble. The distribution of values is not homogeneous; the limestone and dolomite have a similar distribution, although the former has more *outliers* (Fig. 1.11). The most homogeneous distribution is that of the marly limestone, but there are only six samples. The average interconnected porosity for the calcarenite, 0.0846, is higher than that of the limestone, 0.0318, which in turn is slightly higher than that of the dolomite (0.0296). The lowest mean values correspond to the marble (0.0094) and marly limestone (0.0223).

The *specific yield* ranged from zero to 0.0798, with an average of 0.00579, a standard deviation of 0.0118 and a coefficient of variation of 203%. During the tests, 79 yielded no water, including the two marble and eight marly limestone samples. Due to this abundance of zero readings, the frequency distribution (Fig. 1.12) shows a major deviation to the left. The highest value is seen in a sample of dolomite, also the rock with the highest average (0.00724). The averages of the calcarenite (0.00640) and limestone (0.00517) are slightly lower.

The dynamics of these tests yields interesting information. To release their gravity water, 11 samples took more than 30 min in the centrifuge, the equivalent of 2–2.5 years of percolation in the natural environment. Five were of limestone, three of calcarenite and just one of dolomite. However, their behaviour is not homogeneous, even for the same lithology. Figure 1.13 shows the dynamics of the drainage of the samples, which is prolonged, demonstrating variation of S_t/S over time, where S_t is the instantaneous specific flow and S the total specific flow. The dolomite samples gave up all their water in 40 min, while the calcarenites took 120 min and the limestone more than 180 min. The pores of calcarenite and dolomite are larger than in limestone and give up about 90% in 30 min, while the limestone took more than 60 min to do the same.

Relative drainability is the quotient of the specific flow and the interconnected porosity; 79 samples have a value of zero. The average is 0.0758, the standard deviation 0.126 and the coefficient of variation 167%. The maximum value (0.666) is seen in a dolomite sample. This lithology has the highest mean value (0.131). The average value of limestone (0.0882) is higher than that of calcarenite (0.0505). The angle of the distribution of calcarenite is greater than that of limestone and dolomite (Fig. 1.13), indicating a narrower range of variation.

The values of the hydraulic conductivity of the samples analysed are between 2.71×10^{-12} and 5.35×10^{-5} m/s, with a geometric mean of 4.25×10^{-10} m/s, a standard deviation of 3.98×10^{-6} and a coefficient of variation of 1230%. The minimum value was obtained in a sample of calcarenite and the maximum in a limestone. The distribution for each type of rock is very similar, apart from the loamy limestones, which have a narrow range. The highest geometric average is seen in the dolomites (5.52×10^{-10}) and limestones (5.10×10^{-10}), while the lowest average corresponds to the marly limestone (3.45×10^{-11}) and marble (1.68×10^{-10}). The geometric mean of the hydraulic conductivity in the calcarenite is 4.19×10^{-10}. This indicates that the hydraulic conductivity of the matrix in carbonate massifs is very low, but that the fractures and fissures—as well as the subsequent processes already indicated—increase the k-value.

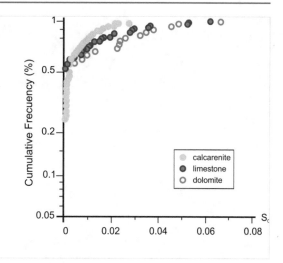

Fig. 1.11 Cumulative frequencies of interconnected porosity

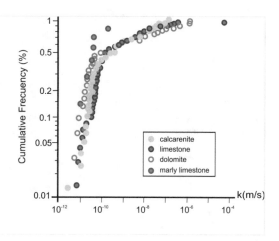

Fig. 1.12 Cumulative frequency of the specific flow rate

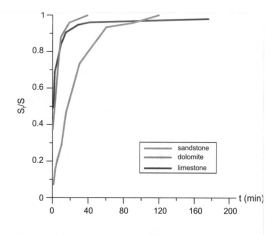

Fig. 1.13 Sandstone drainage dynamics for samples that yield water over a prolonged period

1.6.4   Box 3: Sector North of Ronda

1.6.4.1   Field Data

Statistical analysis of fracturing allows it to be applied to specific problems. The sector studied covers an approximate surface area of 60 km^2 to the north of Ronda (Malaga; Fig. 1.14). From a geological point of view, it is at the western end of the Betic Cordillera, the most internal sector of the external zones.

The lithostratigraphic series begins with Muschelkalk limestone, of an approximate thickness of 20 m, followed by the Keuper marl (50 m), the object of the present study. This is a thick carbonate formation (500 m) of mainly Jurassic age, overlaid with Cretaceous marl and facies flysch sediments of highly variable thicknesses. The series ends with a discordant deposit of Miocene detrital material (molasses). The apparently simple geometry of the folds is complicated by overthrust and retro-overthusts, as well as a very dense network of fractures (Fig. 1.15). This is confined to the external zone.

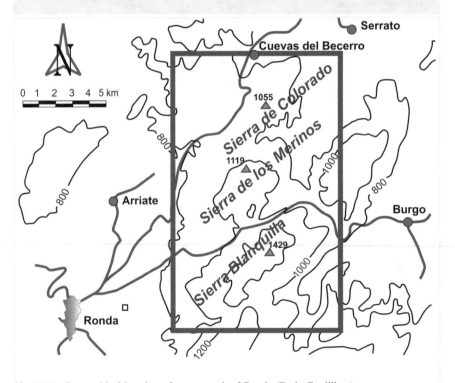

Fig. 1.14 Geographical location of sector north of Ronda (Betic Cordillera)

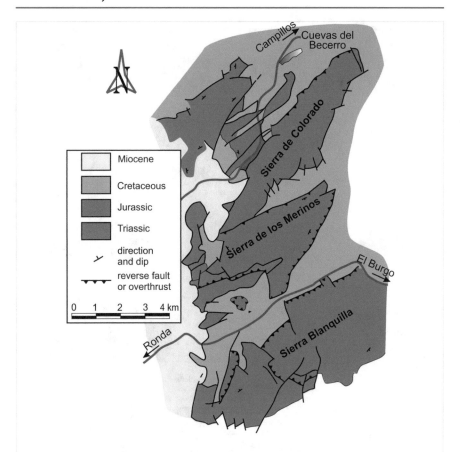

Fig. 1.15 Geological scheme. The Jurassic is essentially made up of carbonate rock

Figure 1.16 shows the values both for the sector overall and grouped by each of the three massifs in question (Sierras Blanquilla, Los Merinos and Colorado; Photos 1.4 and 1.5). By each massif, the diagrams have a more homogeneous aspect than overall, a logical consequence of the increased number of fractures, clearly seen in the general diagram. In Sierra Blanquilla, southern sector, there are two strong trends: 140°–160° and 50°–70°, while in Sierra Colorado, northern sector, fractures running 20°–30° and 90°–100° predominate in a less obvious way. In the central sector, Sierra de los Merinos, a certain superposition of effects may be observed, showing a greater dispersion due to the many relative maxima and minima. The general diagram presents a homogeneous aspect, with a slight trend (6.3%) around 150°–160° and another (6.1%) 20°–40°. Fractures aligned between 50° and 90°, together with the trends noted, comprise more than 40% of the total.

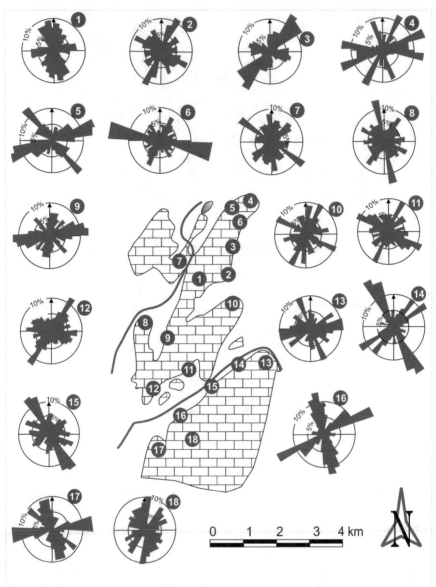

Fig. 1.16 Fracture measurements (field data)

Photo 1.4 Sierra de las Nieves, from Burgo (Photo M. T. Leboeuf)

Photo 1.5 Detail of Sierra Blanquilla (Photo M. T. Leboeuf)

1.6.4.2 Analysis of Aerial Photo

A scale of approximately 1/25,000 was used with stereoscopic pairs. By tracing the fractures identified onto an overlay, we constructed a fracture map (Fig. 1.17) as described by the Optical Bank of the Hydrogeology Laboratory of the USTL (Montpellier, [33]). Scanning a high-contrast version with a laser beam enabled us to create directional frequency diagrams, expressed as percentages of the cumulative length of fractures of the total of the area, for each of the three sectors studied. This technique was later replaced by digital counting.

In the grouped field data, the fracturing appears more homogeneous than in the fracture map in Fig. 1.18, due to the larger number of elements considered. The directional distribution is, however, quite different. In Sierra

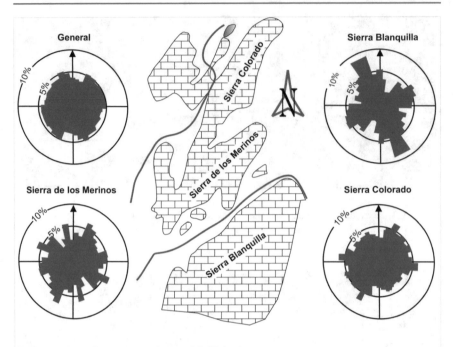

Fig. 1.17 Fracturing measures (grouped field data)

Blanquilla, a strong trend of 140°–160° is clearly seen in the microfracture diagram, but this is not the case with the second strongest trend seen on the fracture map (20°–40°), which is barely perceptible in the field data. In Sierra Colorada, the strongest pattern apparent on the aerial fracture map is again at 140°–160°, yet this has no correspondence to field sampling, while the trend of 90°–100° very well represented in the microfracture maps yet is insignificant in the photogeology mapping. In the Sierra de los Merinos, the fairly isotropic distribution is reminiscent of that in the field data.

Finally, the overall fracture map shows trends different from those seen in the field, with a predominance of fractures with a marked northerly component; more than 45% are aligned 30°–40°. This demonstrates a discrepancy between the results obtained by the two methods described. One explanation is the aforementioned poor representativeness of the sampling. Another is the methodology used in each case, since the field results were expressed as a percentage of fractures in terms of their number, while those of the aerial photo were expressed as a percentage of their cumulative length. This difference gives rise to the phenomenon of 'discrimination' against certain fractures, when grouped in 'families' of 10° intervals across the four sectors.

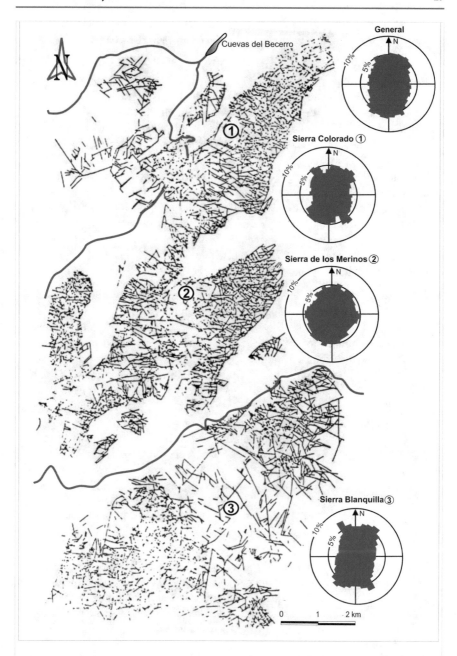

Fig. 1.18 Overall fracture map

1.6.4.3   Partial Analysis by Sampling the Fracture Map

Distribution of Fracture Lengths

Results were obtained from a sample of this fracture map, comprising 754 fractures. The parameters concern their directional distribution on the basis of both number and percentage of cumulative length, their density and mean length of each 10° 'family' of fractures. The values are therefore to be regarded as an approximation. Figure 1.19 shows the sectors chosen for this purpose.

Graphical representation of the data shows their good fit to a log-normal distribution. The approximate mean length of the fracture families is esti-mated at 220 m, with extreme values of 270 m (30°–40° and 50°–60°) and 170 m (130°–140°); the standard deviation is 28 m and the coefficient of variation 0.12. The variation in the approximate magnitude of this parameter is appropriate to the scale, and its distribution clearly reflects the random nature of the phenomenon that caused it.

Directional Distribution of Fractures; Fracture Intensity

Values for the distribution of frequencies of families of fractures (by number and cumulative length) were obtained from the same data of 754 fractures, grouped in families of 10° across the four sectors, as mentioned. For methodological reasons, the count was carried out using circular surfaces with progressively increasing radii (250, 500 and 1000 m, termed Levels 1, 2 and 3, respectively), apart from Sector D, where only radii of 1000 m were used. This led to three distributions of fracture intensity, in terms of number of fractures). Figure 1.20 shows the results and the phenomenon of buffering the peak alignment by increasing the number of fractures analysed, as is logical. The tendency was to move closer and closer to our general Optical Bank fracture map.

If each sector is considered separately, by increasing the sampling area, the variations in the fracture distribution measures can be studied. These results were compared to those of the field fracture analysis. For this purpose, Sectors A, B and C focused on other field measuring stations (Nos. 2, 9 and 15, respectively). Figure 1.21 shows the values.

It can be observed how in Sector A the main trends (20°–30°, 70°–80° and 100°–120°) appear as the sampling area increases, yet their relative frequency decreases as a logical consequence of the increase in the number of fractures considered. It is interesting to note that one of the above patterns barely perceived in the field, namely 70°–80°, becomes increasingly important rel-ative to the other two.

Similarly, the 20°–30° pattern decreases noticeably when the sampling surface is increased. In Sector B, as the magnitude of the strongest trend becomes relatively less, the diagram's shape does not undergo much varia-tion. It is worth noting that the progressive development of strong alignment to the 140°–150° interval, acceptably represented in the field diagrams, is

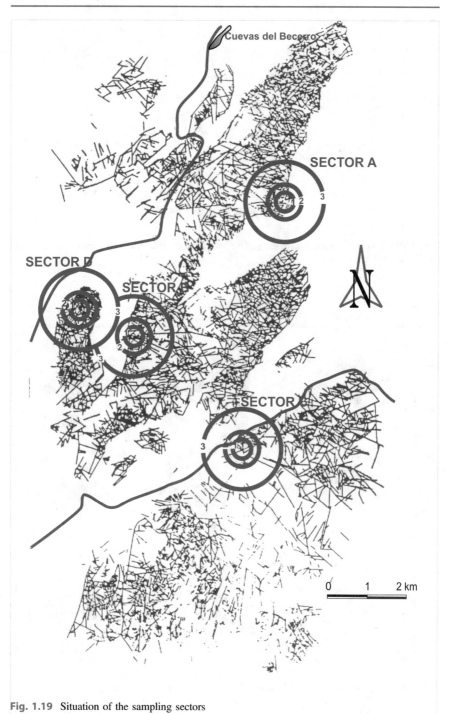

Fig. 1.19 Situation of the sampling sectors

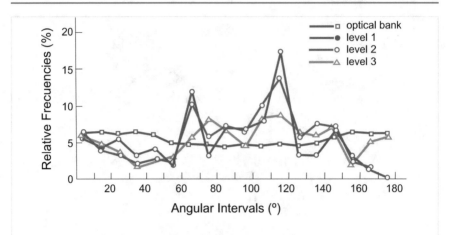

Fig. 1.20 Results of the analysis of each of the levels and comparison to the general values on the Optical Bank fracture map

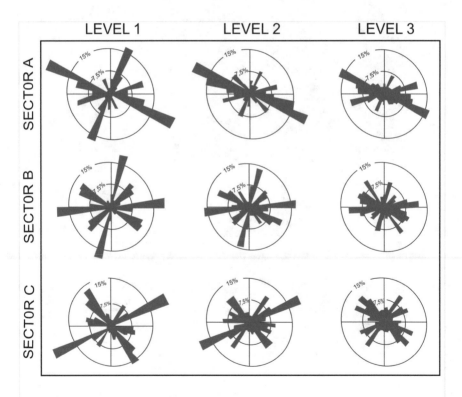

Fig. 1.21 Variation in fracture intensity (FL) in each sector for each level

barely perceptible in the first sampling level. In Sector C, also the main trends in the intervals are retained, yet when the sampled surface is increased, the 60°–70° interval decreases notably.

Figure 1.22 presents the values of mean lengths (LM) and cumulative lengths (LA) of fracture families, obtained from sampling as a whole. The values of the two distributions, in relative frequencies expressed as percentages for fracture intensity, are also shown: FN refers to the total number of fractures (754), while FL refers to the total length, approximately 168 km. The two distributions show a great similarity, as evidenced by the high degree of linear fit achieved, with a correlation coefficient of 0.94. We studied the minor differences between the corresponding values of both distributions (D) to determine whether there is a specific reason. The linear regression of D on LM was tested, and the value of the correlation coefficient (0.93) shows the goodness of the fit of the adjustment. The line equation is $D = 4.7272 - 0.0209$ LM.

The absolute value of the deviation is minimum if the mean length of the family coincides with the global mean value of the total fractures, 226 m, which is the value of the abscissa from the origin. Likewise, the extreme absolute values of D will appear in intervals whose mean lengths are furthest from the overall mean value (M). In our case, as $D = FN - FL$, the positive values of D will correspond to families with mean lengths of less than M, while the negative ones will be at intervals with an LM parameter value greater than the general mean. These considerations are a rough study of the phenomenon of the 'discrimination' against certain families of fractures arising from the methodology used.

Fracture Density

The variation in the distribution of fracture density due to the methodological reasons noted is of a specific type, since it refers to the four sectors where sampling was carried out by aerial photo. The values obtained, expressed in both the number of fractures/surface and length of fractures/surface, are shown in a fracture density map (Fig. 1.23) for the three areas considered.

On the basis of this map, the global average values of fracture density for the total area investigated, based on aerial photo at 1/25,000 scale, were found to be 150 fractures/km^2 and 30 km of fractures/km^2. From these values, the approximate total number on the fracture map has been estimated, as well as their corresponding cumulative length, to be 7500 fractures, at this scale totalling 1500 km in length. Although representative of the scale of work at which they have been carried out, these values should be regarded as indicative. If the observation scale is changed, the results are different yet have identical statistical parameters.

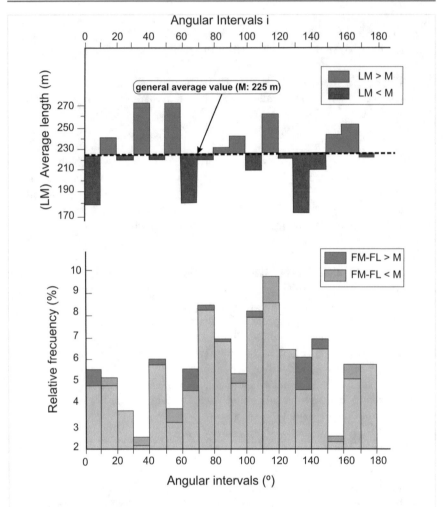

Fig. 1.22 Mean length of fracture families (LM) in relation to the overall mean value (*M*), and graphical representation of the two distributions obtained (FN and FL) for the fracture intensity parameter

1.6.5   Box 4: Sorbas Gypsum

To the east of the town of Sorbas (Almería), there is a strong set of material, essentially of Messinian gypsum, in which a dense and spectacular karst network has developed. Although the tectonic activity affecting the area was never violent, the aim was to determine the influence, if any, that its fracturing had on the genesis of the large number of karst caves present. Given that surface observation conditions are unfavourable and that few fractures are visible in aerial photo, fractures were measured in two quarries (Photo 1.6 and 1.7) where gypsum has been exploited, taking the precaution of

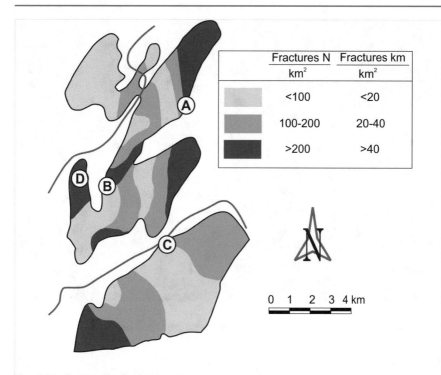

Fractures N	Fractures km
km²	km²
<100	<20
100-200	20-40
>200	>40

0 1 2 3 4 km

Fig. 1.23 Fracture density map

measuring only very clear instances whose origin is unrelated to the drill holes used in mining (Photo 1.8 and 1.9).

It was possible to measure 67 fractures, most of them subvertical. The results, grouped in intervals of 15°, are shown in Fig. 1.24. It is observed that the most frequent interval (29.8%) lies 151°–165°, and that, 70.1% of fractures are between 136° and 180°.

If we compare these results with those obtained for the direction of rectilinear cave fragments (17 topographies from 17 caves were used, collected in four speleological studies) across 63 measurements, it can be seen that the strongest is slightly displaced; in fact, the most frequent direction is the interval 136°–150° (22.2%; Fig. 1.25); 49.2% of these caves have an orientation of 136°–180°.

If we take into account the lithological nature of the material in question, it seems logical to conclude that there is a marked tectonic influence on the caves, although other factors also play a role.

For comparison to the previous data, the directions of 23 rectilinear channel fragments and other alignments that are potentially assimilated into fractures were also measured. It was observed that the strongest alignments were unlike each other (26.1% in the interval 46°–60°), although there is a

Photo 1.6 View of the Sorbas gypsum, showing several gypsum quarries

Photo 1.7 Detail from one of the gypsum quarries (Photo A. Pulido)

Photo 1.8 Quarry faces of Sorbas gypsum (Photo A. Pulido)

Photo 1.9 Quarry faces of Sorbas gypsum, a detail (Photo A. Pulido)

Fig. 1.24 Histogram of the
fractures measured in the
quarries. $N = 67$

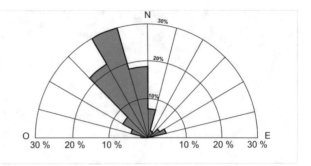

Fig. 1.25 Histogram of
rectilinear cavity fragments.
$N = 63$

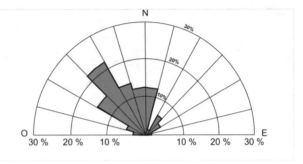

Fig. 1.26 Histogram of
direction of channels and
other alignments in gypsum
(Aerial photo)

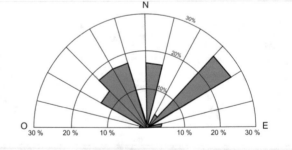

relative maximum corresponding to one previously noted (Fig. 1.26). This
disparity may well be because the drainage network is predominantly influenced by factors other than fracturing, such as lithological contact and the
inclination of layers. In sum, it is concluded that there is a remarkable
structural control of the rectilinear sections of caves; if we correct the field
measurements of the magnetic declination, the strongest alignments are
acceptably coincident.

1.6.6   Box 5: Sierra Gorda

1.6.6.1   General Treatment

Sierra Gorda is one of the largest carbonate massifs in the Betic Cordilleras (293 km^2). It is almost entirely of limestone and dolomitic material from the Middle Triassic, belonging to two stratigraphic and tectonic units: the Sierra Gorda unit and the Zafarraya unit. The actual thickness of the carbonate series is unknown, but exceeds a thousand metres (Photo 1.10 and 1.11). The tectonic deformation of the region is both polyphasic and highly complex.

As a whole, the Sierra Gorda unit is a large, elongated dome running in a north–south direction, the result of the superposition of two folding systems with directional axes of 150° and 30°–50°. The latter correspond to a first folding phase, probably associated with a certain northwest displacement of the entire unit. These folds are, in general, of great radius, open and more visible in the northern sector of the massif thanks to the highly visible stratification of the Liassic limestone there. A second folding phase caused 150° asymmetrical folds facing southwest, whose inverted flanks are considerably shorter than the normal flanks. Within these synclines, the highest of the stratigraphic series (Upper-Middle Jurassic and Cretaceous) is often preserved.

1.6.6.2   Fracture Analysis

The first step in any fracture analysis of aerial photo consists of drawing up a fracture map at a scale of 1:18.000. Using the RAFRAC programme, this identified 42,352 traces of fractures [34]. This programme consists of several subprogrammes, two of which were used: RAFNUM numbers the fracture field, while RAFSET calculates the quantitative elementary parameters identifying the fracture (orientation, length and surface density) and performs statistical tests (not geostatistical). The data procedures considered both the orientation of the fractures (in terms of both number and cumulative length) and their length.

Figure 1.27 shows the directional distribution of the fractures by the number of fractures (a) and their cumulative length (b) across the whole carbonate massif, as well as separately for both the Sierra Gorda and Zafarraya units. The strong similarity between the histograms is striking, corresponding to the two modes of expression of the intensity of fracturing, indicating in a qualitative way the high correlation between the values expressed by the number and those by the cumulative length of fractures within each directional family. Razack's studies on the relationship between these modes of expression, using statistical methods of linear regression, conclude that it is linear and positive, so that fracture intensity parameter can be expressed in either one way or another, regardless. The linear correlation coefficients are high for the subsectors into which Sierra Gorda was subdivided by the study, ranging from 0.69 to 0.97, with a 95% confidence

Photo 1.10   Sierra Gorda, from the surroundings of the flooded Zafarraya polje (Photo A. Pulido)

Photo 1.11   Sierra Gorda, from the surroundings of the flooded Zafarraya polje (Photo A. Pulido)

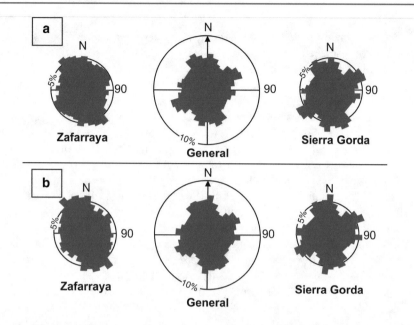

Fig. 1.27 Directional distribution of fractures (**a** number of fractures; **b** cumulative length) in the Sierra Gorda and Zafarraya units, also of the set of both (general), from aerial photo data

interval. The lowest values systematically correspond to those subsectors with the fewest data.

The circular histograms that group the (general) fracture orientation data in Fig. 1.27a, b are essential to our understanding of the Sierra Gorda aquifer's fissure structure. Close observation reveals the main families of fractures present in the massif:

- Fractures oriented north–south to 0°–10°. In general, these are long fractures (175 m is medium length).
- Fractures oriented 50°–70°. These constitute a well-defined and very uniform alignment in terms of number of fractures and cumulative length.
- Fractures oriented 90°–100°. Although not very frequently seen in aerial photo data, they are usually large (average length is 183 m).
- Fractures oriented 140°–150°. They are very abundant, although not especially long. They can be easily confused with the north–south fractures.

In the Sierra Gorda unit, these directional trends remain identical, or are perhaps even more strongly defined, which shows their preponderance (29,904 fractures) over the entire massif. The Zafarraya unit (12,304 fractures) presents certain differences: on the one hand, the north–south alignment is present, characterized here by large fractures; the alignment of 50°–70° E is unclearly displayed and is widely dispersed; the alignment to 90°–100° is tenuous, both in terms of number of fractures and cumulative length; an alignment to 140°–150° is observed, passing gradually towards north–south, without either being clearly distinct. In this unit, 140°–150° fractures are predominant over those of other families.

Figure 1.28 shows the fracture intensity values in terms of cumulative length for 16 subsectors across the massif on the basis of geological and fracture density criteria. A wide spatial variability is observed from one sector to another in the position of the directional alignment, as well as the order of its relative magnitude, in all probability due to the presence of large structures influencing this distribution locally. However, the variability is not as great as that obtained by field measurements, with some overlap of adjacent sectors close (S–F and S–H, for example). The passage from one to another is often characterized by the appearance or disappearance of one of the fracturing alignments (S–B, S–A, S–H). The similarity to the general diagram shown in Fig. 1.30b grows with the number of fractures in that particular sector.

With respect to the length of the fractures, the software identified their distribution across 20 intervals of 50 m. The graphic representation for the fracture map as a whole can be seen in Fig. 1.29. The histogram is unimodal and has positive asymmetry. The peak is in the range of 100–150 m. This resembles log-normal distribution. In the case of Sierra Gorda, the graphical representation of the fracture length data showed a good fit to a log-normal probabilistic scale. The average length of fractures is about 165 m, with extremes of 147 m (40°–50°) and 183° (90°–100°). The standard deviation is 0.106 m and the coefficient of variation is 6.5.

The analysis of the tectonic microstructures makes it possible to specify the nature of the data supplied by the aerial photo. A microtectonic study at specific measuring stations showed an almost general absence of stylolithic structures in the aquifer, probably due to the great purity of the carbonate materials, forcing us to consider the fault striations and the planes that contained them mainly as kinematic markers, as well as tension diaclasses with calcite fillers.

This study carried out 114 measurements of fault planes and corresponding grooving at nine field stations across the Sierra Gorda massif. Where it could be perceived, any extension movement was noted. Most of the stations were in newly excavated quarries or road or railway cuttings and tunnels; measurements at each were taken in specific and precise sectors of very short extent to avoid measuring areas that had experienced varying ellipsoid stress.

1.6.6.3   Preferential Direction of Karstification

To compare fracture tectonics and the development of karstification, a statistical study was carried out on the direction of the galleries in known karst caves in the Sierra Gorda sector. The method consists of drawing the network

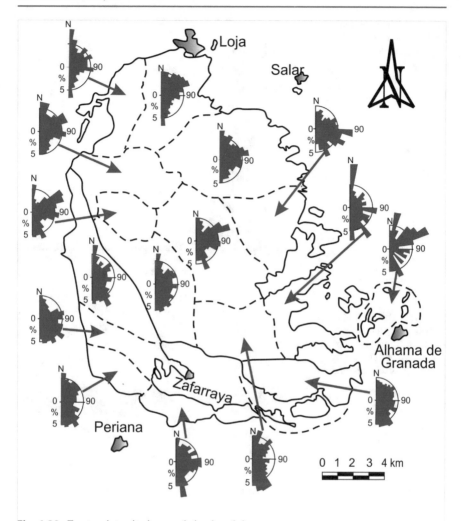

Fig. 1.28 Fracture intensity in cumulative length by sector

of galleries, rectilinear sections or axes of galleries, especially those with horizontal development, on the caves' speleological topographical plan. This involved careful observation, paying special attention to the appearance of the transversal sections, the contours of the cavity and the presence or otherwise of secondary formations (stalactite deposits, sandy sediments, collapses, etc.). Spiral or meandering steps and large collapsed caverns were not considered. It was necessary to orient each plan to the geographic north to discount the magnetic declination on its date of construction.

The length and direction of 219 straight sections of the corresponding established karst networks were measured from the topography recorded in

Fig. 1.29 Fracture length distribution for the fracture map as a whole

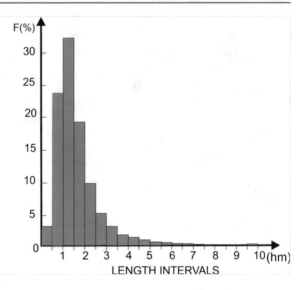

Fig. 1.30 Distribution of the known karst galleries in Sierra Gorda: **a** By gallery number; and **b** By their cumulative length by directional family (B)

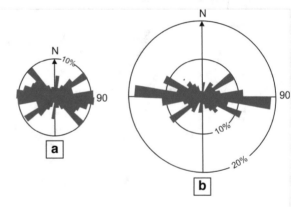

77 caves. The total cumulative length of the set of rectilinear sections is 2920.7 m, giving a very small theoretical average density of this karst network, of the order of 10 m of galleries per km^2, taken across the whole surface of the karst (293 km^2). Nevertheless, a statistical study was carried out, taking into account directional point data and also the alignment and cumulative length of galleries, representing the data in circular histograms with a class interval of 10° (Fig. 1.30).

The characteristics of the caves used in the statistical analysis are irregularly distributed across the massif. Most are in the Lower and Middle Liassic limestone of the Sierra Gorda unit, in a central band-oriented east-west. The average altitude of their entrances is 1325 m asl, with a standard deviation of 182 m. The maximum penetrated depth is 149 m, recorded in the Sima de los

Machos. No cavity reaches the aquifer's water table, and there are neither perennial nor seasonal groundwater courses, apart from a penetrated stream-sink in the Zafarraya polje; thus, it is an inactive network. Vertical caves in 1 m mid-opening diaclasses, sinkholes and caves dominate, favouring certain stratification, yet always showing a marked structural influence.

The results of statistical analysis of the directions and cumulative length of the karst galleries (Fig. 1.30) show a split into several alignments, as follows:

- 0°–10°: the peak is slightly marked but clearly separate.
- 50°–6° E: is a relatively distinct alignment. It shows great homogeneity in terms of equilibrium between the number of galleries (Fig. 1.30a) and their cumulative length (Fig. 1.30b).
- 80°–100°: in the histogram of cumulative lengths shows a very marked maximum, the system being the longest penetrable relative length.
- 110°–140°: the angular dispersion is very large, and there is no strong correlation between the histogram of cumulative lengths and the histogram of number of galleries. Our opinion is that it corresponds, essentially, to 140° but that it shows a deviation towards the east-west position, with extensive galleries running in that direction.

1.6.6.4 Fracturing and Karstification

When comparing the distribution histograms of fractures from aerial photo (general histograms in Fig. 1.28) to the orientation distribution histograms derived from the karst galleries (Fig. 1.30), there is almost perfect correspondence. The following considerations can be made on a case-by-case basis:

- Comparison of the fracture intensity diagram expressed in numbers of fractures (Fig. 1.30a) to the karstification intensity diagram expressed in numbers of karstification galleries (Fig. 1.30): the fracture aligned 0°–10° shows a certain karstification, but it is not the dominant one; the 50°–70° fractures have developed intense karstification, as have the 90°–100° and 140°–150° fractures. The deviation of the karstification alignment from the fracturing alignment is minimal.
- Comparison of the diagram of fracture intensity expressed in cumulative lengths of fractures (Fig. 1.30b, general diagram) to the karstification intensity diagram expressed in cumulative lengths of karst galleries (Fig. 1.30b): the fractures aligned 0°–10° present a very low degree of karstification; fractures 50°–70° show a well-defined peak of karstification intensity; fractures running 90°–100° constitute a family that is weakly defined regarding its intensity of fracturing yet shows the most extensive karstification. The fracture system running 140°–150° does not show

remarkable karstification, as large galleries have not developed, but the most developed are aligned 110°–120°. An explanation for this development of karstification is difficult; perhaps, it is because some fractures running 140°–150° were inflected towards east-west, and it is these that became karstified. On the other hand, it might correspond to fractures in the east-west system that were somewhat displaced (10°) from the trend.

From all that has been said so far, it must be noted that the maximum karstification is presented by the fractures running 90°–100°. It is an inactive karst network, so it has been developed in the past. Its coincidence with 90°–100° fractures, the oldest fault system created during the alpine orogeny, leads us to ask at what time this system acted extensionally. It seems that, in the last compressive stages of the orogeny, the maximum axis of stress lay approximately E-W, which favored the opening of fractures of similar orientation. This situation persisted until the end of the Miocene, when a new phase of deformation produced the Mio–Pliocene discordance within a framework of a new stress system.

During the Upper Miocene, the 140°–150° and 50°–70° fracture systems were already established. The second of the two was probably the most extensional, and its formation resulted in the closure of the first, the 140°–150° fractures. During the last alpine compressive stages, both systems were susceptible to karstification, but the maximum karstification would still be in the east-west direction (maximum distension). The morphological features of the massif, as well as the nature of the materials deposited in the Granada depression (lacustrine limestone), support the idea of this karstification in the Upper Miocene.

There is no evidence of how karstification has developed in the aquifer since the end of the Miocene. The stress system proposed for this period and up to the quaternary seems to indicate a tendency to open up fractures running 140°–150° E and north–south, and less intense in those 90°–100° and 50°–70°. It is probable that the karst network was progressively embedded during the Pliocene and during the quaternary by the progressive uplift of the massif, promoting the most open fractures, in two or more stages of maximum intensity of karstification. If one considers the young age of the relief of this sector of the Betic Cordillera, it is logical to regard the current karst network as still undergoing development.

A key aspect of fracturing is its at-depth evolution, especially in terms of density and possible 'closure'. Numerous studies undertaken in quarry cuttings postulate a logarithmic decrease in depth, which would mean that, a few metres down, the number of fractures and their openings is rapidly reduced. There would be a particularly porous decompressed surface strip, which would quickly close up at depth. Data obtained by means of boreholes contradict this conclusion; although it may be realistic in a non-karst fissure environment, it should be discarded as an approximation of karst. To this we must be added the observation that this superficial strip is the most vulnerable to the impervious effect of the insoluble fraction, the clay component and/or the alterites that may fill discontinuities. Moreover, this applies to both a karst medium and a fissured one.

Another reflection of practical interest is the real application of such fracture analyses to—for example—the siting of boreholes. Although this is discussed in Chap. 6, we can advance that conscientious studies of fracturing usually have less weight than simple criteria, for example, ease of access, the altitude of the selected site and ownership of the land, among others. However, fracturing analyses have remarkable practical application and are indispensable in civil works, where groundwater may affect their progress [35, 36]. They are also of interest in mining [37, 38].

1.6.6.5   A Unique Example: Torcal de Antequera

Torcal de Antequera is a unique karst massif of great beauty (Photos 1.12, 1.13, 1.14 and 1.15 [39]). Jurassic limestone and dolomite make up several anticlines, as illustrated in the cross sections in Figs. 1.31 and 1.32. Aerial photo were subjected to a detailed analysis of the area's fracturing in Chap. 2 [five], and results are summarized in Fig. 1.33. The circular histograms show the fractures by sector [40], alongside one giving the overall picture. A generality in this type of analysis is that sectoral results do not have to coincide with the overall pattern; in fact, it is more usual for them not to coincide, indicating that distribution is heavily influenced by local phenomena.

Photo 1.12 Uniqueness of Torcal de Antequera is portrayed in this image (Photo A. Pulido)

Photo 1.13 Uniqueness of Torcal de Antequera is portrayed in this second image (Photo A. Pulido)

Photo 1.14 Uniqueness of Torcal de Antequera is portrayed in this third image (Photo A. Pulido)

Photo 1.15 Uniqueness of Torcal de Antequera is portrayed in this fourth image (Photo A. Pulido)

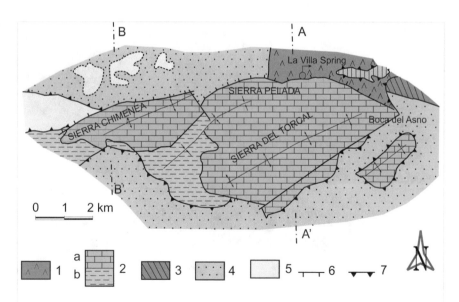

Fig. 1.31 Hydrogeological map of Torcal de Antequera: (1) Trias of Antequera; (2) **a** Torcal unit; **b** Cretaceous marl; (3) Sierra de las Cabras unit; (4) Aguila complex; Flysch unit; (5) Post-orogenic Miocene; (6) Normal fault; (7) Overthrusts A-A' and B-B': representative cross sections

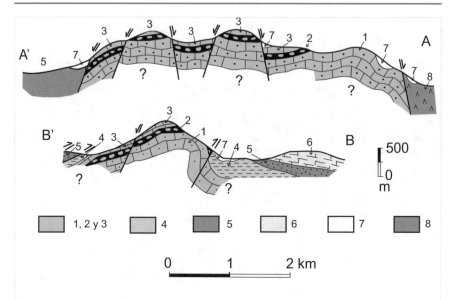

Fig. 1.32 Hydrogeological cross sections of localities in Fig. 1.31: (1) Oolitic limestone; (2) Nodular limestone; (3) Brecciated dolomite; (4) Marl; (5) Flysch; (6) Miocene; (7) Hillside debris; (8) Antequera Triassic

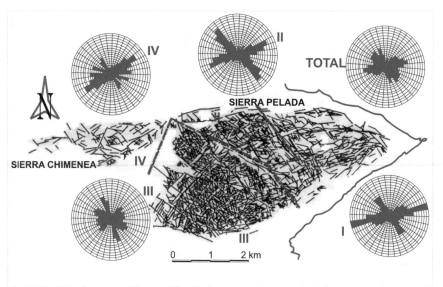

Fig. 1.33 Main fractures of the massif, with frequency histograms both by sector and as a whole

1.7 Further Reading

Bertrand, C., Denimal, S., Steinmann, M., Renard, P. Eds. 2019. *Eurokarst 2018, Besançon: Advances in the Hydrogeology of Karst and Carbonate Reservoirs* (Advances in Karst Science). Springer, 249 pp.

Fidelibus, M.D., Pulido-Bosch, A. Eds. 2019. Inland and coastal karst aquifers: Functioning, monitoring and management. *Geosciences,* Special issue ISSN 2076-3263. 10 pp.

Ford, D.C., Williams, P.W. 2007. *Karst Geomorphology and Hydrology,* 2nd edn. London: Unwin Hyman, 601 pp.

LaMoreaux, P.E., Wilson, B.M., Memon, B.A. 1984. Guide to the hydrology of carbonate rocks. *Studies and Reports in Hydrology,* 41.

Mudry, J., Zwahlen, F., Bertrand, C., LaMoreaux, J.W. Eds. 2014. *H2Karst Research in Limestone Hydrogeology.* Springer, 190 pp.

Parise, M., Gabrovsek, F., Kaufmann, G., Ravbar, N. Eds. 2018. Advances in karst research: Theory, fieldwork and applications. *GSL Special Publications,* 486 pp.

Pulido-Bosch, A., Motyka, J., Pulido-Leboeuf, P., Borczak, S. 2004. Matrix hydrodynamic properties of carbonate rocks from the Betic Cordillera (Spain). *Hydrological Processes*, 18: 2893–2906.

Renard, P., Bertrand, C. Eds. 2017 Advances in the hydrogeology of karst and carbonate reservoirs. *EuroKarst 2016, Neuchâtel* (Advances in Karst Science). Springer, 370 pp.

Younos, T., Schreiber, M., Kosič Ficco, K. Eds. 2018. Karst water environment. *Advances in Research, Management and Policy.* Springer, 275 pp.

1.8 Short Questions

1. How important are karst aquifers to the supply of urban drinking water?
2. How many meanings are there to the concept of karst?
3. How is pseudokarst defined?
4. What is the meaning of thermokarst material? Give examples.
5. How many types of karst terrain do you know?
6. What are the conditions for the formation of karst in quartzite rock? Give examples.
7. What is *carniolan rock* and how can it be generated?
8. What are fluid inclusions? Describe what you know.
9. How many types of porosity are recognized in a karst environment?
10. Why is effective porosity usually only a small fraction of the total porosity in a limestone or dolomite?
11. What role does matrix porosity play in a karst aquifer?
12. Briefly explain how you would perform a fracture analysis in a karst massif.
13. To what extent does fracturing determine the hydraulic parameters of a karst massif?
14. Why is there a great disparity between the distribution of fracture frequencies in the data obtained from aerial photo and in situ measurements?
15. To what extent do you think fracturing influences karstification? Which fractures are promoted most?

1.9   Personal Work

1. Karst and pseudokarst terrain and its characterization.
2. Types of porosity, identification and quantification in karst terrain.
3. Assess the extent of karst terrain fracturing from aerial and satellite images.
4. Assess the extent of karst terrain fracturing from on-the-ground observations.

References

1. Ford, D. C., & Williams, P. W. (1989). *Karst geomorphology and hydrology* (2nd ed., 2007, p. 601). London: Unwin Hyman.
2. Williams, P. W. (Ed.) (1993). Karst Terrains. Environmental changes and human impact. *Catena Supplement, 25*, 268.
3. White, W. B. (1988). *Geomorphology and hydrology of karst terrains* (p. 464). New York: Oxford University Press.
4. Pulido-Bosch, A. (1996). Los acuíferos kársticos españoles. *Investigación y Ciencia, 232*, 50–56.
5. Llopis, N. (1970). Fundamentos de Hidrogeología kárstica. In Blume (Ed.), *Fundamentos de Hidrogeología kárstica* (p. 269). Madrid.
6. Plan BLEU. (2004). *L'eau des méditerranéens: Situation et perspectives* (PNUE/PAM. MAP Technical Report Series No. 158) (p. 366). Athènes.
7. Ayala, F. J., et al. (1986). *Memoria del Mapa del Karst de España* (p. 68). IGME.
8. Avias, J., & Dubertret, L. (1975). *Phénomènes karstiques dans les roches non carbonatées* (pp. 31–40). París: I.A.H.
9. Hill, C. A., & Forti, P. (1986). *Cave minerals of the world* (2nd ed. 1997, p. 463). Huntsville, USA: National Speleological Society.
10. Castany, G. (1984). Hydrogeological features of carbonate rocks. In: *Guide to the hydrology of carbonate rocks* (pp. 47–67). Studies and Reports in Hydrology, UNESCO.
11. Motyka, J. (1988). *Triassic carbonate sediments of Olkusz-Zawiercie ore-bearing district as an aquifer* (p. 109). Cracovia: Tesis University.
12. Martín, J. M. & Pulido-Bosch, A. (1981). Consideraciones sobre la porosidad y la 1126 permeabilidad en dolomías. *I Simp. Agua en Andalucía*, I, 337–346 (Granada).
13. Martín, J. (1980). *Las dolomías de las Cordilleras Béticas* (p. 201). Tesis University of Granada.
14. Pulido-Bosch, A. (1985). L'exploitation minière de l'eau dans l'aquifère de la sierra de 1058 Crevillente et ses alentours (Alicante, Espagne). In *XVIII Congrès Hidrogéol. Intern.* (pp. 142–149). Cambridge.
15. Pulido-Bosch, A., Morell, I., & Andreu, J. M. (1996). Hydrogeochemical effects of 1061 groundwater mining of the Sierra de Crevillente Aquifer (Alicante, Spain). *Environmental Geology, 26*, 232–239.
16. Zuber, A., & Motyka, J. (1994). Matrix porosity as the most important parameter of fissured rocks for solute transport at large scale. *Journal of Hydrology, 158*, 19–46.
17. Atkinson, T. C. (1977). Diffuse flow and conduit flow in limestone terrain in the Mendip Hills, Somerst (Great Britain). *Journal of Hydrology, 35*, 93–110.
18. Mangin, A. (1975). *Contribution à l'étude hydrodynamique des aquifères karstiques* (Thèse Doct.) In *Ann. Spéléol.* (pp. 29–3, 283–332, 29–4, 495–601, 30–1, 21–124).
19. Bakalowicz, M. (1986). On the hydrogeology in karstology. *Jornadas sobre el Karst en Euskadi, 2*, 105–129.
20. Małoszewski, P., & Zuber, A. (1992). On the calibration and validation of mathematical models for the interpretation of tracer experiments in groundwater. *Advances in Water Resources, 15*, 47–62.

21. Maloszewski, P. (1994). Mathematical modelling of tracer experiments in fissured rocks. *Freiburger Schriften zur Hydrologie, 2,* 1–107.

22. Motyka, J., Pulido-Bosch, A., Borczak, S., & Gisbert, J. (1998). Matrix hydrogeological properties of Devonian carbonate rocks of Olkusz (southern Poland). *Journal of Hydrology, 211,* 140–150.

23. Motyka, J., Pulido-Bosch, A., Pulido-Leboeuf, P., & Borczak, S. (2002). Propiedades hidrogeológicas de la matriz de rocas carbonatadas de la Cordillera Bética (sur de España). *Geogaceta, 32,* 311–314.

24. Pulido-Bosch, A., Motyka, J., Pulido-Leboeuf, P., & Borczak, S. (2004). Matrix hydrodynamic properties of carbonate rocks from the Betic Cordillera (Spain). *Hydrological Processes, 18,* 2893–2906.

25. Kiraly, L. (1975). Rapport sur l'état actuel des connaissances dans le domaine des caractères physiques des roches karstiques. In A. Burger & L. Dubertret (Eds.), *Hydrogeology of karstic terrains, international union geological science series B* (Vol. 3, pp. 53–67). Kenilworth, UK: IAH.

26. Rats, M. V., & Chernyasnov, S. N. (1965). Statistical aspect of problem on the permeability of the joint rocks. *Actes Colloque de Dubrovnik* (Vol. 1). AIHS.

27. Benavente, J., Fernández Gutiérrez del Alamo, R., Fernández-Rubio, R., & Pulido-Bosch, A. (1980). Algunas consideraciones metodológicas para el estudio de la fracturación en acuíferos kársticos. *Actas Espeleológicas, M-1,* 19–54.

28. López Chicano, M. (1992). *Contribución al conocimiento del sistema hidrogeológico kárstico de Sierra Gorda y su entorno (Granada y Málaga)* (p. 429). Tesis Doct., University of Granada.

29. López Chicano, M., & Pulido-Bosch, A. (1993). The fracturing in the Sierra Gorda karstic system (Granada). In A. Pulido-Bosch (Ed.), *Some Spanish karstic aquifers* (pp. 95–116).

30. Pulido-Bosch, A. (1982). Consideraciones hidrogeológicas sobre los yesos de Sorbas (Almería). *Reun. Mon. sobre el Karst Larra, 82,* 258–274.

31. Borczak, S., Motyka, J., & Pulido-Bosch, A. (1990). The hydrogeological properties of the matrix of the chalk in the Lublin coal basin (southeast Poland). *Hydrological Sciences Journal, 5,* 523–534.

32. Prill, R. C., Johnson, A. J., & Morris, D. A. (1965). *Specific yield laboratory experiment showing the effect of time on column drainage* (1662 B). Geological Survey Water-Supply Paper.

33. Drogue, C., Mas, G., Grillot, J. C., Lloria, C., & Guérin, R. (1975). Studies on light filtration and fracturing of limestone rocks (hydrogeology). *Revue de Géographie Physique et de Géologie Dynamique, 17*(1), 39–44.

34. Razack, M. (1986). *Application de méthodes numériques et statistiques à l'identification des réservoirs fissurés carbonatés en Hydrogéologie* (p. 384). Tesis Doct., University of Montpellier.

35. Milanovic, P. T. (1981). *Karst hydrogeology* (p. 434). Colorado: Water Resources Publications.

36. Bonacci, O. (1987). *Karst hydrology. With special reference to the Dinaric Karst* (p. 184). Berlín: Springer-Verlag.

37. Younger, P. L., Banwart, S. A., & Hedin, R. S. (2002). *Mine Water Hydrology Environmental Pollution, 5,* 127–270.

38. http://www.imwa.info/

39. Sanz de Galdeano, C. (1993). Principal geological characteristic of the Betic Cordillera. In A. Pulido-Bosch (Ed.), *Some Spanish karstic aquifers* (pp. 1–7).

40. Fernández–Rubio, R., Jorquera, A., Martín, R., Zofio, J., Villalobos, M., & Pulido-Bosch, A. (1981). Análisis de la fracturación y directrices estructurales en el acuífero kárstico de El Torcal de Antequera (Málaga). *I Simposio Agua en Andalucía, II,* 659–673 (Granada).

Karstification and Forms

2

2.1 Glossary

Karstification works in two ways: one *chemical*, involving corrosion; and the other *mechanical*, involving physical erosion. The first dominates initially, but later the second may become dominant.

The kinetics of *dolomite dissolution* is slower than that of calcite, although the solubility of dolomite is greater. The higher the Mg content of water, the longer the transit time of that water within the aquifer, and the same goes for other parameters.

The *hypersoluble* rocks showing the most karst forms are evaporites, gypsum and halite. The dominant process is dissolution, which starts with halite; even at room temperature, up to 350 g/l of NaCl can be dissolved in water.

The action of erosion on a rock surface is inversely proportional to the thickness of the overlying soil and the density of the vegetation cover, both factors closely related to climate.

The most abundant *insoluble residue* is clay, which due to its iron content has a red colouring, hence the name *terra rossa*. This clay can clog fissures and ducts.

Karst processes are *erosion, corrosion* and *incrustation*, both superficial and exokarst, underground and endokarst, with others between these domains.

Exokarst forms include: *karren; dolines*—whether funnel, trough, window, tectonic or structural; *uvalas; poljes*—formed at conjunction of *dolines* or tectonic areas, by contact or structural; *valleys*—both blind and dead-end; *canyons; water losses*, sinkholes or *ponor*; and *chasms*—caused by sinking or collapse, by the excavation of the sump, by ascending or inverse erosion, or by a combination of such processes.

© Springer Nature Switzerland AG 2021
A. Pulido-Bosch, *Principles of Karst Hydrogeology*, Springer Textbooks in Earth Sciences, Geography and Environment,
https://doi.org/10.1007/978-3-030-55370-8_2

Endokarst forms are conduits and networks of conduits between caverns—essentially they are horizontal developments, including *corridors, galleries* and *caverns*. Within these caves, we find forms of erosion-corrosion, such as **forced conduits** and *giant's cauldrons*; *clastic block cones*, *block rubble* and *detritus terraces*, and also of reconstruction, such as *stalactites, stalagmites, eccentrics, columns, flags, towers, microgours, cornices* and *cave pearls*.

The strangest superficial forms are *gypsum 'tumuli'*, circular or elliptical bulges ranging in size from a few centimetres to more than 5 m along the major axis, with or without collapsed dome.

2.2   Karstification

Karstification involves two major groups of processes: one is essentially *chemical* —corrosion—and the other is more *mechanical* and physical-erosion. Initially, it is the first that dominates but, as *conduits* are created, erosion becomes more relevant and perhaps more extensive than corrosion.

2.2.1   Corrosion

This term refers to the physicochemical processes by which rock is dissolved and transformed to another state. The solubility of $CaCO_3$ in pure water is very low, in the order of 14–15 mg/l at 25 °C and partial pressures of CO_2 in normal conditions (0.0003 atmospheres); when the partial CO_2 pressure increases, the solubility rises considerably. The $CaCO_3$ solution comes about in the essential presence of water by chemical attack by an acid (nitric, nitrous, organic, sulphuric and so on) and, above all, by CO_2, the key component of its aggression. This attack leads to a change of phase from solid to liquid, governed by the law of equilibrium of Gibbs' phases: there is equilibrium when the chemical potential of one phase (solution) is equal to the potential of the other (solid). In non-equilibrium, a component will spontaneously change to the phase of greater potential and, in the case of karst, this means that corrosion will occur when the potential of the liquid phase is greater than that of the solid [1].

In the presence of CO_2, HCO_3 carbonic acid is created in the water. This attacks the limestone, creating calcium bicarbonate; and this simplified reaction is reversible, so $CaCO_3$ can be both dissolved and precipitated:

$$CO_2 + H_2O + CaCO_3 (\Leftrightarrow HCO_3)_2$$

In reality, the process is far more complex, since it is a system of heterogeneous equilibrium with reactants distributed over three phases (gas, liquid and solid), in each of which is an equilibrium chain [2]. Any modification to one of the partial equilibria entails a readjustment to the others. The solution's equilibrium is governed by three interdependent variables: (CO_2), Ca^{2+} and H^+, constant for a given temperature provided there are no complications. When the proportions are disturbed, we are faced with either precipitating or aggressive water. It must be borne in mind that the reaction speeds are very different, so the system evolves at the speed of the slowest.

$CaCO_3$ is much more soluble in an acid than an alkali. The pH/Ca^{2+} equilibrium, expressed in mg/l of CO_3Ca with calcite at 15 °C, is 500 at pH 6.5 and 42 at 8.5 (700 and 60 in aragonite at 10 °C). In P CO_2 it is 60 mg/l CO_3Ca at 10^{-3}, and 450 at 10^{-1} for aragonite at 15 °C. These parameters of aggression and precipitation depend on, among others, the acidity, the free CO_2 and Ca^{2+}, also the temperature and the ionic strength.

In general, the likelihood of incrustation is higher in an alkali; a decrease in P $_{CO2}$ will cause this, too, although it is a function of the concentration of the various ions in water. However, many authors have found that karst spring water is mostly not at saturation point and is aggressive. Solutions that are oversaturated in CO_3Ca are metastable; that is, CO_3Ca does not precipitate immediately. There are other compounds that intervene, both actively and through their catalytic action, in the processes of both chemical attack and dissolution. The most relevant may be: the H_2SO_4 formed by oxidation of the sulphides present in the rock, among other causes; nitric acid (probably created by storms, and in the soil as a result of the actions of certain bacteria); and organic acids (humic acid in soil) from the decomposition of organic matter.

These are complex processes, involving geological, physicochemical, climatic, biochemical and hydrological factors. How do they take place in the natural environment? The presence of so many factors entails successive and frequent switches between aggressive and precipitating water. Given the scale involved, this results in the dissolution of large masses by minor variations in the concentration of $CaCO_3$ between oversaturation and subsaturation. Regarding those **aspects related to CO_2**, while this gas is present in the air it is produced especially in the subsoil. It moves out of the air by diffusion and from the latter in both gaseous form—either by diffusion or gravity, as it is denser than air—and in solution, at the rate of 2.15 l/m^3 at 0 °C and 0.8 l/m^3 at 25 °C. Between the soil and the saturated zone, therefore, there is an atmosphere rich in CO_2 in hollows and fissures that, if in contact with air, will be more impoverished than one that is isolated from such effects. $CaCO_3$ is usually deposited in ventilated channels of circulating groundwater. At depth (in the saturated zone), the depletion of CO_2 due to temperature decrease lends a greater aggression to the water. Mixing several waters of dissimilar concentration usually results in an aggressive water [3].

Temperature influences the process of karstification, since the solubility of carbonates normally increases slightly with temperature. In the presence of CO_2, the opposite is true: at lower temperatures, the solubility of CO_2 increases. Additionally, at higher temperatures the equilibrium reactions are faster, while the CO_2 content in soils decreases as the temperature drops, due to reduced biological activity.

Thus, temperature acts in dissimilar ways on the equilibrium equations, sometimes in contradictory ways. By one means or another, the effects of temperature can be compensated for and, for this reason, karst developments are present in a range of glacial, Mediterranean and tropical climates [4], and in the last, the denudation is far greater since the level of biological activity is so high.

In terms of the corrosion of magnesium carbonates, magnesite is more soluble than calcite, at 94–117 mg/l at 25 °C and normal P CO_2, and much higher if it is $MgCO_3$, hydrated or nesquehonite, $MgCO_3.H_2O$. Dolomite's main constituent is $CaMg(CO_3)_2$. By virtue of the common-ion effect, the solubility should be lower as it should multiply the solubility products of Ca^{2+} and Mg^{2+}, which are inferior to the unit. Experience shows that karst waters become saturated sooner in calcite than in dolomite; further, the kinetics of dolomite dissolution is slower than that of calcite. Moreover, dolomite rock's lithological and structural characteristics, among other factors, tend to inhibit its dissolution.

2.2.2 Dissolution of Hypersoluble Rocks

The hypersoluble rocks that most commonly develop karst forms are evaporites, gypsum and, above all, halite. In all cases, the dominant process is dissolution, which starts with halite. Water can dissolve up to 350 g/l of NaCl at room temperature, so this material is usually not present as an outcrop, except in arid climates, although the sinking phenomena associated with its dissolution may be detected at the surface. Next comes gypsum, at a solubility of 2.5 g/l at room temperature. Its karstification is very fast, although it may become buffered or vanish quickly at depth. There are many examples around the world, and Spain has spectacular gypsum caves and surface phenomena [5].

2.2.3 Erosion

This section deals with erosion in its broad sense, including meteorization (wind action, sudden changes in temperature, ice, thawing, root action) and mechanical action, both with and without the intervention of water (gravity action, with detachments and classic processes). Erosion acts at the same time as corrosion in karstification process and, similarly, depends on many factors. It takes place both on the surface and at depth. The vital ingredient in corrosion is water, which has a great potential energy that is transformed into kinetics as it meets the ground and circulates through the subsoil.

The action of erosion on a rock surface, in usual climatic conditions, is inversely proportional to the thickness of the overlying soil and the density of the vegetation cover, both closely related to climate. Meteorization of the rock, causing superficial cracking, when accompanied by intense rainfall produces a considerable depth of deposits in low-lying areas, normally with a clay matrix. This explains the presence of saturated detrital beds in the great karst forms of temperate latitudes.

The erosion of the karstifiable mass itself is especially noticeable along open fractures, along which water may circulate on a massive scale. This type of erosion loses its effectiveness a few metres below the saturated zone. Its reverse, a zone of upward shaft development formed by ascending erosion, may develop due to high water pressure beneath the saturated zone.

There is a stage in the karstification of the unsaturated zone, when, along open subvertical fractures—whether caused by tectonics or corrosion—erosion becomes more active than corrosion. Its effect is similar to that of river water and is a consequence of the kinetic energy of the combined mass of water and the material that it transports, both in suspension and dragged along, continually impacting on the floors and walls. In short, at a certain point mechanical action becomes more important than corrosion, actively contributing to hollowing out caves in certain parts, selectively. Given that the mechanical action of water depends on its flow, speed and mass transported, in equilibrium with its kinetic energy, this contributes to widening the conduits, which in turn allows greater circulation and, therefore, greater erosive power. This introduces a *hierarchy of* karstification that is not general but selective, since not all discontinuities are exploited—just certain parts. It arises as a consequence of both mechanical action and corrosion, which is also selective.

2.2.4 Insoluble Substances

Were the limestones, dolomites and other rocks in karst areas completely pure, the entire mass would disappear without leaving any insoluble residues. The reality is very different since, as was mentioned, many insoluble or less soluble substances accompany them. The most abundant insoluble residue is clay, the study of which can provide data on the sedimentation medium and its stratigraphy. This residue has red colouration due to its iron content, hence the generalized term of *terra rossa*. Because it is deposited where conditions are favourable, filling any fissures and conduits, by reducing the permeability this clay has hydrogeological importance. Moreover, when there is an abundance of decalcification clay, one of the best ways to develop a borehole is by means of introducing a polyphosphate dispersal agent into the water.

Another insoluble residue, or one with very low solubility, is quartz, or silica. This is generally not abundant. However, there is a remarkable hydrogeological aspect of the Milky Quartz that is widespread in the Cretaceous calcarenite and limestone series of the external Prebetic Zone. The quartz is in the form of grains the size of sand or larger, and it causes problems in drilling boreholes. The borehole may collapse or, if the water needs to be pumped up, the grains may be sucked into the pumping gear and damage it. The presence of Milky Quartz must be taken into account and a gravel filtering medium incorporated into the cavity.

All these processes lie behind the formation of karst aquifers (Fig. 2.1) which, due to their complexity, give rise to highly diverse typologies, as will be seen in the following chapter.

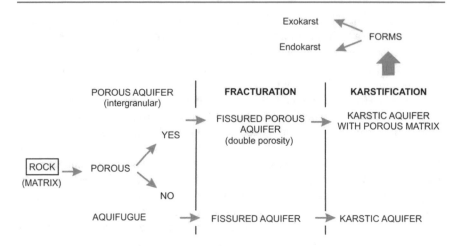

Fig. 2.1 Process sequence in karstification

2.3 Forms Resulting from Corrosion and Incrustation

2.3.1 Main Forms

Various forms of corrosion and incrustation are studied in the field of karst geomorphology. Many are of great hydrogeological interest, and the classifications are diverse, some already classic in Chap. 1 [5], such as the concept of karst apparatus. This encompasses the various forms through which water flows, from running over and under the ground's surface to emerge in a spring: absorption; conduction; and emission. This has its limitations, so the most usual classification considers superficial or exokarst forms, and underground or endokarst forms, with some forms in transition between the two.

Included in the *exokarsts* are: *karren* (Photo 2.1) of highly diverse types and sizes; *dolines* (Photo 2.2)—in the form of a funnel, trough or window, and both tectonic and structural; *uvalas* and *poljes* (Photos 2.3 and 2.4)—at a conjunction of either dolines or tectonic areas, and both through contact and structural; *valleys*—both 'blind', or with no continuation from a given point, and 'dead', without any flow and now located above the currently active bed; *canyons*; *losses*, *sinkholes* or *ponor* (Photos 2.5a,b); and *sinkholes* formed by either sinking or collapse, by excavation of a sump, by ascending or inverse erosion, or by a combination of these processes.

Endokarst forms par excellence are *conduits* and networks of conduits; and *caverns*—with essentially horizontal development, including *corridors*, *galleries* and *caverns*. Within such cavities, we find forms of erosion-corrosion, such as *forced conduits* and '*giant's cauldrons*'; classic *block cones*, *block rubble* and

Photo 2.1 Karren near the doline in Photo 2.2

Photo 2.2 Spectacular doline NE of Castell de Castells (Alicante)

detrital terraces. We also find reconstruction: *stalactites, stalagmites, eccentrics, columns, flags, casts, towers, microgours, cornices* and *cave pearls,* to mention the most common, although more have been described.

Photo 2.3 Zafarraya's *polje* on the southern edge of Sierra Gorda (Granada)

Photo 2.4 Satellite image of Sierra Gorda

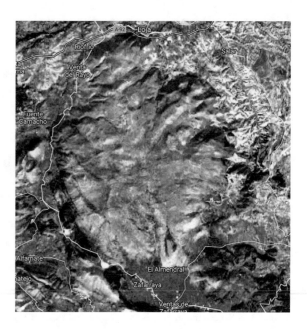

Sinks, chasms, **sinkholes** and the various conduits are of major hydrogeological importance, as they permit rapid flow and/or ingress of large quantities of water from the surface to the aquifer's interior.

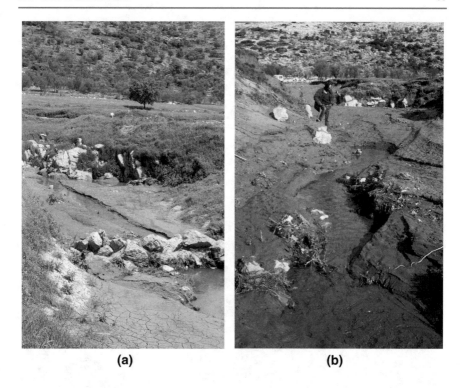

(a)   (b)

Photo 2.5   a,b. One of the *ponor* of the polje de Zafarraya, eastern end, northern edge (Photos A. Pulido)

2.4   Box 1: Sorbas Karst

To the east of Sorbas, there is a set of strata essentially of **Messinian gypsum** corresponding to the Yesares member of the Caños formation (spatial distribution shown in Fig. 2.2). These gypsums are intensely karstified and contain possibly the largest gypsum network, by area, in the entire world. In the River Aguas series [6], 12 cycles of gypsum material alternate with pelitic-carbonate laminated sediments. Some of the gypsum cycles are 20 m thick, making a total thickness of 130 m, and the bands are differentiated: 37 m of the gypsum is pelleted. This is a selenite gypsum, with large crystals of primary origin. The area is gently folded, with a large-radius syncline visible in the gypsum—gypsum is more consistent than the underlying detrital terrain.

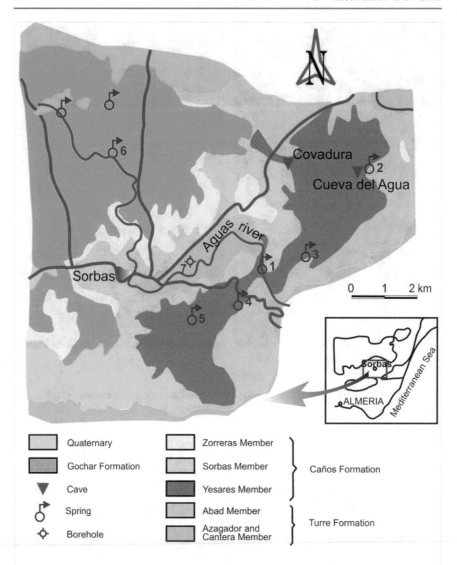

Fig. 2.2 Geological scheme of the Sorbas gypsum and its surroundings

2.4.1 The Karst Forms

1. The Exokarst

The Sorbas gypsum occupies only 14 km^2 at the surface, with a maximum length in a NNE direction of 10 km and a maximum width of 4 km. The River Aguas has excavated a spectacular gorge in crossing the outcrop. Given the high solubility of gypsum, at 2.6 g/l, considerably greater in the presence of NaCl, both exo- and endokarst forms are abundant and, in many cases, unique.

Of the exokarst forms, dolines and small poljes dominate, the latter with a highly irregular geometry usually filled with silts and clays. One of the large dolines is about 50 m in diameter, with subvertical walls, and is interpreted as evidence of a *sinkhole* relating to the network of underlying ducts. An underground complex of caves, some related to the Cueva del Agua, starts from this doline. Lapiés with acute crests of a centimetre high can be seen here.

The most peculiar surface shapes are the *gypsum tumuli,* the circular or elliptical domes ranging from a few centimetres in size to more than 5 m along their major axis, with or without a collapsed dome (Photo 2.6). An explanation for each phase of their development is given in Fig. 2.3. In the first phase, a superficial hollow occurs in the first 10 centimetres or so of a gypsum layer, also a carbonate medium. Although it must be a less resistant plane (a), this does not necessarily coincide with a stratification surface. Weathering leads to an increase in volume and the uplift of that layer. Subsequent research [7, 8] has allowed us to go deeper into this lifting process, and it is apparent that the intercrystalline dissolution and the rapid new precipitation, with water mediation, create a further increase in volume and the consequent bulge (b). Continued weathering ends up with the dome

Photo 2.6 One of the numerous gypsum tumuli visible on the surface of the Sorbas massif (Photo A. Pulido)

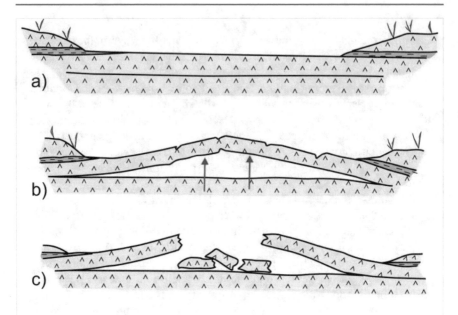

Fig. 2.3 Stages in the genesis of a perforated gypsum tumulus

losing its mechanical resistance, leading to its eventual collapse (c). The
shapes visible on the ground represent Phases (b) and (c).

A statistical study of 81 of these forms [9] showed 47 with a collapsed
dome, 33 circular forms and 48 elliptical forms. The following parameters
were measured (Fig. 2.4): the circles' diameter; both axes of elliptical forms;
the height of the bulge; and the thickness of the raised gypsum layer. The
diameters ranged from 3.6 m down to 0.2 m, with a mean value of 1.56 m
and a standard deviation of 1.17. For the elliptical shapes, the major axis
(a) ranged from 11.70 m down to 0.2 m. The average large axis was 1.78 m,
with a standard deviation of 1.96. The minor axis (b) ranged from 13.2 m
down to 0.1 m. These values have a log-normal distribution, like so many
other parameters in nature. The bulge of the gypsum tumuli, calculated as the
quotient (a–b)/a, is of an average of 0.15 m with a standard deviation of 0.16.

Fig. 2.4 Diagrammatic cross section of a gypsum tumulus with a perforated dome, showing two
measured parameters (*e* and *h*)

Regarding the circular gypsum tumuli, 43.6% of the measured shapes had a bulge of up to 0.1 m, and only 10% exceed 0.4 m. Details of the formation of this bulge are shown in Fig. 2.5.

The relationship was also determined between the thickness e of the convex layer and the height h of the gypsum tumulus' perforated dome. The maximum of e was 0.5 m. The average was 0.1 m, with a standard deviation of 0.10 m. The height h of the gypsum tumulus, measured on the external face ranged from 1.35 m down to 0.33 m, with a standard deviation of 0.28 m, among 37 individuals. The statistical relationships between the parameters measured in the population are indicated in Table 2.1.

2. The Endokarst

Sinkholes are common across the massif, and it is a peculiarity that many have a fig tree within them, taking advantage of the higher humidity there. They are chasms of a variety of forms, although the tubular ones dominate, either rounded or elliptical, and of a fairly constant vertical cross section. These forms are in transition between exo- and endokarst and allow access to

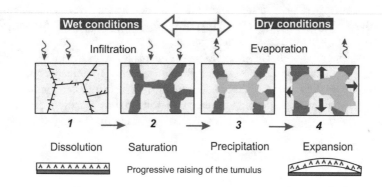

Fig. 2.5 Successive processes that raise a gypsum tumulus (Adapted from [8])

Table 2.1 Equations of regression lines

Parameter		n	Linear equation	r
h	a	37	$h = 0.18\,a + 0.14$	0.79
	a*	36	$h = 0.29\,a + 0.96$	0.90
	b	37	$h = 0.27\,b + 0.14$	0.77
	b*	36	$h = 0.35\,b + 0.06$	0.94
	e	36	$h = 2.20\,b + 0.09$	0.90
e	*a*	46	$e = 0.09\,a + 0.01$	0.93
	b	46	$e = 0.12\,b + 0.00$	0.95

*Equation after eliminating an anomalous value; n: pairs of values; r: correlation coefficient

numerous caves. Many of them are interconnected, as recent topography studies have shown, forming large complexes. Among them are Agua Cave and the Covadura complex, each measuring about 10 km. Both have water in the lower levels—that is, they descend to the phreatic level—and display a similar genetic scheme that Calaforra has dubbed 'the vadose erosion of marly interbedded strata'. This consists of vertical wells that correspond to gypsum strata, and almost horizontal sections developed in the pelitic inter-calations, with a roof formed of a gypsum stratum with evidence of proto-conduits. Parietal deposits, stalactites and stalagmites are common, together with other forms described for the first time in this area. Sometimes, the flows that infiltrate these complexes are very high, eroding metapelite materials and leading to *caverns* of considerable size, causing their roofs to collapse.

At present, more than a thousand caverns are known in the Sorbas gypsum karst, representing almost 100 km of underground galleries, of which 8.6 km are in the largest cave in Spain, developed in gypsum: the Cueva del Agua. This massif is vulnerable due to intensive gypsum mining since the 1980s [10, 11]. Of the areas with a high density of karstification and numerous interconnected caves, the following stand out:

The Cueva del Agua System, also called Cueva del Marchalico, is in the northern sector of the gypsum outcrop and is the largest cavern complex in Andalusia. There are currently 24 known *sinkholes* (Fig. 2.6) that allow access to the network. It has a subterranean watercourse that is the source of the Las Viñicas spring. The great doline forming the cavity was possibly an ancient polje that evolved into a great karst depression with numerous dolines inside. The system runs through the first two levels of the gypsiferous series, of about 50 m thick, descending through one of the loamy *marl* intercalations until it reaches the impermeable level marked by the underlying Tortonian *marls*.

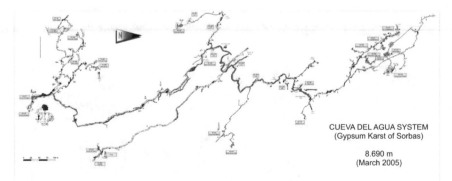

CUEVA DEL AGUA SYSTEM
(Gypsum Karst of Sorbas)

8.690 m
(March 2005)

Fig. 2.6 Topography of the Sorbas water cave (*Source* EspeleoClub Almería)

The Covadura System is in the northern sector of the gypsum outcrop, adjacent to the mining area. This is one of the deepest of all Spain's gypsum caves, reaching a depth of 120 m and with more than 4 km of galleries. It has been excavated through the gypsum series in such a way that all the inter-calations, marl levels and interstratified gypsum are picked out on the walls of its galleries [12]. At its bottommost point, the piezometric level of the aquifer varies between a depth of 120 and 105 m. In the 'Gallery of the Forest', the Covadura System features hollow stalagmites, unique the world over [13]. Other relevant caves are the Corral System, the Sima del Campamento (at 130 m, the deepest in Sorbas) or the Sima del Yoyo and Sistema B-1. They all have a genesis and configuration of levels similar to that of the Covadura System and have caving passages longer than 1 km.

The KAS System is immediately south of the Covadura area, and it represents one of the most intense karst areas in Sorbas gypsum, with features such as Deer Cave, Siphons Cave, Noise Cave, Stalagmites Cave and the Kass Cave Complex, all interconnected by water.

The GEP Complex is to the north of the depression of Cueva del Agua, with seven known entrances. It is developed on three gypsum levels and has more than 1 km of galleries. Communication has not been ruled out between this system and other nearby systems, such as the Treasure Area or the Water Cave Area. The GEP Complex has perhaps the greatest variety of spe-leothems in the entire gypsiferous outcrop [14].

The Cueva del Yeso and the Barranco del Infierno are in the southern sector of the Sorbas outcrop. This is the underground part of the Infierno ravine route. It has more than 1 km of galleries, with continuous water cir-culation. Its origin is attributable to the fluvial capture of the ravine by a cavity initially developed at a lower gypsiferous level. It has four known entrances, two of vast dimensions at level of the watercourse. Tesoro, in the southern sector of the gypsum outcrop, is an iconic feature of the Sorbas karst as it has a cave with the greatest morphological richness of any gypsum karst. The Cueva del Tesoro (Treasure Cave) has a passage of almost 2 km, with meanders, huge caverns with stalactite formations and gypsum clusters of 2 m in length, as well as evidence of a prehistoric human presence. There is a watercourse running through the last part of this system. Other relevant caves in this sector are the Cueva de los Apas, the Sumidero Baena and the Sima del Estadio. The last has the largest known underground cavern of gypsum karst, at almost 2000 m^2, caused by a major collapse of the upper gypsum stratum.

The Cueva del Peral, in the ravine of the same name, is adjacent to the Cueva del Agua and the Cueva de los Apas, in a sector where the epikarst system stores more water. The Cueva del Peral is its most important cave, with almost 2 km of galleries and a good number of lakes along the way.

Some unique forms in these caves are hollow gypsum, *hole stalagmites*, present especially in the Covadura Forest gallery. Strange *gypsum balls* appear by exudation on the walls of some important Sorbas caves (Cueva del

Agua, Cueva del Tesoro and Covadura, among others). The extensive gypsum arrowhead crystal twinning is especially notable in the Cueva del Tesoro (Galería de los Cuchillos, with hundreds of medium-sized 50 cm ×20 cm gypsum crystals; and Galería de los Cristales, with crystals longer than 2 m) and is practically unique. Gypsum tray stalactites are another beautiful feature of Covadura and similar caves.

2.4.2 Hydrogeological Aspects

For various reasons, gypsum is not usually considered to be of hydrogeological interest. The main drawback is the poor quality of its water, which is unsuitable for most applications, whether for irrigation or human consumption. Also, when associated with halite, the saline content may be as high as that of brine and, if it comes into contact with other limestone karst aquifers, it can cause elevated hydrogeochemical anomalies [15, 16]. A spectacular example of an aquifer in gypsum is that of Jezireh in Syria, which has an area of about 8000 km^2 and springs producing water that is measured in m^3/s—and some wells have very high specific flows of up to 80 l/s/m.

It is clear that these gypsums display the aquiferous behaviour shown by the deep parts of caves and numerous existing aquifers, including the Viñicas spring, which was even used as a water supply. These springs—at Las Viñicas, Yeso cave and Peral spring, although they may have much higher peak flow—generally have a flow rate of less than 1 l/s.

The main discharge of the Sorbas system is the spring in the bed of the River Aguas and in the gorge excavated in the Yesares member, known as Nacimiento de los Molinos de río Aguas (Source of the Mills of the River Aguas), from the ubiquitous mills here. The impervious bottom of the aquifer is constituted of the marl and sandy marl of the Abad member; there is a hydraulic connection with the limestone of the Canteras member along its northern edge. There is a possible lateral continuity of the gypsum towards the west, as shown by boreholes, 42 of them in saturated gypsum. One is 50 m deep and has yielded 80 l/s with an 11 m decline, although it is highly vulnerable to the level dropping in a dry year. The hydraulic gradient between this point and the mainspring of the Nacimiento de los Molinos de río Aguas is 1.3%.

The measurements in the mainspring, carried out in the 1970s by Carulla [17] for over more than a year, yielded flows of between 60 and 175 l/s, estimated by him to average 120 l/s. The ability of this spring to keep flowing has always been surprising, maintaining a flow of more than 50 l/s even in extreme droughts, which indicates that condensation of water vapour in its labyrinthine network of galleries and sinkholes may be a notable element in recharging this system. Although some measurements were undertaken to seek support for this type of contribution, they were insufficiently extensive and no verified records are available to put a reasonable figure on the value of such a contribution.

It should be noted that the drainage basis at the point of emergence is wider than 300 km², so its contribution to the recharge of the system cannot be completely ruled out—whether through alluvial or other permeable points —potentially connecting back into the gypsum.

2.5   Box 2: Gypsum Karst of Vallada (Valencia)

Vallada is a unique example linked to the Keuper facies Triassic evaporite materials in the southwest of the province of Valencia in the municipalities of Vallada, Montesa and Canals. Within the Sierra Grossa, it is a Prebetic alignment that extends from La Font de la Figuera and the Vall de Valldigna and Gandía.

The Triassic outcrop, which includes gypsum thicknesses greater than 100 m, constitutes a strip of just over 3 km oriented southwest–northeast and surrounded by carbonate materials from the Upper Cretaceous (Creu formation). The Miocene materials are represented in conglomerate and bioclastic sandstones that reach up to 40 m in thickness and in the blue and white marl that rests on the previous set and that can also reach great thicknesses. Finally, there are locally well-developed Quaternary slope deposits and decalcification clay. This area shows a remarkable tectonic complexity. The Keuper Trias is apparent along a large fracture (Fig. 2.7). Existing folds have a northern vergence and are thrust over the Miocene marl and the plastic material of the Keuper.

This is a good example of karst in gypsum, with numerous surface and subterranean forms [18, 19]. Within the former, there is abundant karren at the centimetre scale (Photo 2.7) and wider, one uvala and two flat-bottomed dolines of about 15 m in diameter. At the southern edge of the intersection of the two dolines that generate the uvala, there is a small open fracture that acts as a sink. The most striking form is a polje that collects the run-off from a basin of about one km², with a flat bottom furrowed by a ravine that disappears into a vertical wall in the swallow hole known as Els Brollaors. This marks the start of a conduit that can be entered, with a dry channel. Behind a 2-m-long tunnel, there is a small cave 15 m long and 8 m high, and parts of the walls are covered in clay and with dragged-clay boluses; the cave ends in a flooded conduit (Fig. 2.8).

The exit from the cavern is through a circular conduit formed under pressure, with a diameter of 1.5 m giving access to another cavern similar to the previous one, also with numerous boulders, many of them of gypsum. In the middle of the cavern, there is a *dejection* cone corresponding to the

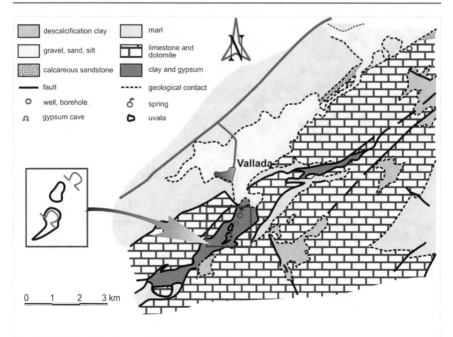

Fig. 2.7 Geological scheme of the Vallada karst area

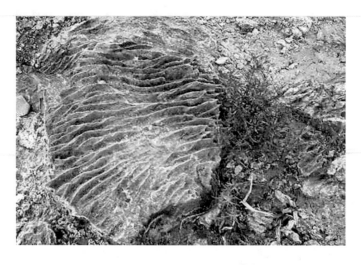

Photo 2.7 Microkarren of sharp ridges in the gypsum of Vallada (Photo A. Pulido)

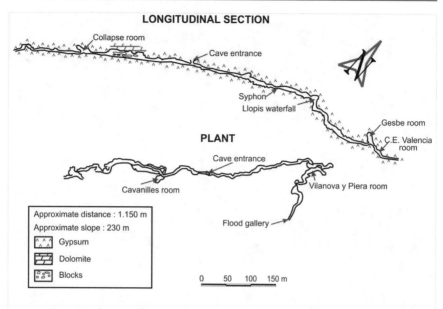

Fig. 2.8 Topography of the Tunnel of Sumidor

shallow uvala. This cavern ends in a flooded and conduit impassible without adequate equipment. The Tunnel of Sumidors is the continuation of Els Brollaors (Photos 2.8a and b), as shown. It is about 1100 m long, and its access point is approximately in the middle. The flow through its interior can exceed 100 l/s in flood periods, but in low-water conditions is only a few litres per second. The interior, which has unstable walls, varies in width from 1 to 5 m, with a predominance of angular shapes and numerous blocks detached from walls and roof (Photo 2.9). Classical processes dominate over dissolution. The dominant constructive forms are calcite (Photo 2.10), forming casts, *pavements* and even *moon-milk* in some open fractures, and there are incipient stalactites.

At about 150 m upstream of the entrance there are elevated galleries, currently inactive, with two caverns of about 12 m wide. The top of the gallery is formed of well-stratified dolomite in an asymmetric anticline (Photo 2.10). A circular gallery more than a metre across leads from its eastern end, with abundant angular dolomitic pebbles forming a cone; this gallery is partially flooded with red clay, and it leads to the Cavanilles cavern. Below the tunnel entrance are two waterfalls, the first of which is 30 m high. At the end of the tunnel is a chamber of about 15 m long by 25 m high (Gesbe chamber). After passing through an area of large blocks, this reaches the final

(a)   **(b)**

Photo 2.8 a,b. Els Brollaors, at the start of the tunnel (2.8a), and the main tunnel entrance (Photos V. Benedito)

Photo 2.9 Detail of the passage, showing the gypsum coated with calcium carbonate deposits (Photo A. Fornes)

Photo 2.10 Triassic dolomite fold within the tunnel, Sala de la Gamba (Photo A. Fornes)

Photo 2.11 La Saraella,
saline water rising up from the
passage floor in a confused
heap of blocks (Photo V.
Benedito)

'siphon', or water tunnel. The water comes from Saraella spring and is highly
saline, as Vilanova indicated [19]. The spring extends a few tens of metres
backward, and it presents a risk of collapse (Photo 2.11).

2.6   Box 3: Pseudokarst Forms on the Southern Edge of Sierra del Maimón (Almería)

The Sierra del Maimón is in the north of the province of Almeria near the town of Vélez Rubio (Figs. 2.9 and 2.10), and it is made up of limestone and Jurassic dolomite [20, 21]. These create an area of notable relief, and an extensive glacis has developed on the upland's slopes [22]. These Jurassic materials 'float' on the more recent marly and sandy formations.

Along the southern edge of the mountainous alignment a wide glacis has developed, comprised of a conglomerate of angular limestone cobbles and dolomite, alternating with clay and red silt, which can locally exceed 40 m thick. At the foot of the Jurassic escarpments, these glacis are covered by alluvial fans and by angular carbonate pebbles of predominantly a centimetre across. In this material numerous closed forms are visible, resembling a typically elliptical doline (Fig. 2.10). Of the 24 identified, 13 currently have surface drainage, whereby the upwelling water has been captured by backward erosion by the ravines.

The largest of these shapes—No. 22 in Fig. 2.11—measures 530 m long. Of the others, seven exceed 100 m and only one is less than 50 m. Elongated shapes dominate, with a major axis/minor axis ratio of between one and 3.1, with an average of 1.6 and a standard deviation of 0.48. We distinguish two groups of these forms: those developed in hillside debris, at an approximate height of 1200 m; and those generated in the glacis. In the former, the walls and the bottoms are of loose material, partially disturbed by human intervention. Closed forms are frequent, staggered and from north to south display decreasing depth.

In the glacis, the forms that have developed are usually of a much greater size, and in some cases, they are authentic *poljes*. In all, the dip of the layers of the glacis depends on whether it the southern or the northern edge, and in general, this is in the opposite direction, indicating rotation. The bottom is

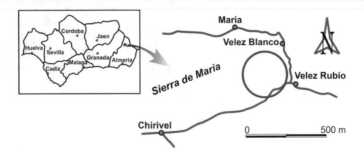

Fig. 2.9 Study area of Sierra del Maimón

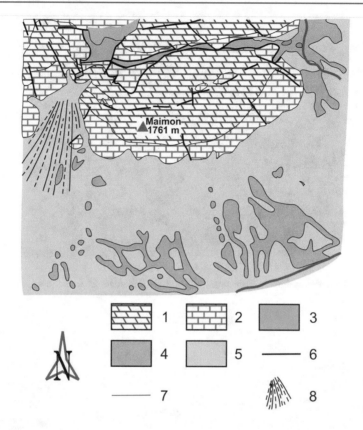

Fig. 2.10 Geological scheme of the environment in which pseudokarst forms develop:
(1) Dolomite; (2) Limestone; (3) Nodular limestone, marlstone and radiolarite; (4) Marl;
(5) Quaternary deposits; (6) Main faults; (7) Geological contact; 8) Alluvial fan

usually either cultivated or planted with pine trees, and there are numerous
angular stones from the glacis itself, within a clay matrix. Most of these forms
are no longer closed, having been captured by erosion from ravines. These
origins lie in the dissimilar effect of the erosion of the Chirivel dry riverbed
from west to east, which is very marked (Figs. 2.12 and 2.13) and creates a
difference in height of about 270 m between the beginning of the glacis and
the floor in the first cross section and 380 m in the second, separated 2.5 km.

The average slope varies from 6 to 15% towards the east, due to the
steepness of the ravine promoting incisive upward erosion into its tributary
beds, which is pronounced on the northern border. These slopes and the
presence of a sandy clay substratum under the glacis material generated
curved plane slides (lystric faults). The break in equilibrium that would ini-
tiate rotational movement would have been provided by the steep slopes

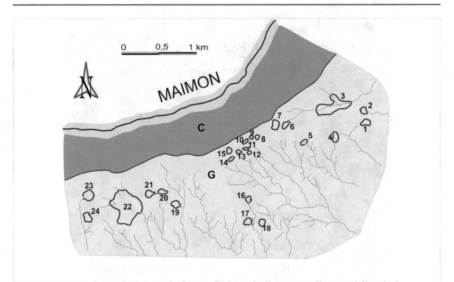

Fig. 2.11 Sierra del Maimón's main forms. C: Jurassic limestone; G: essentially glacis

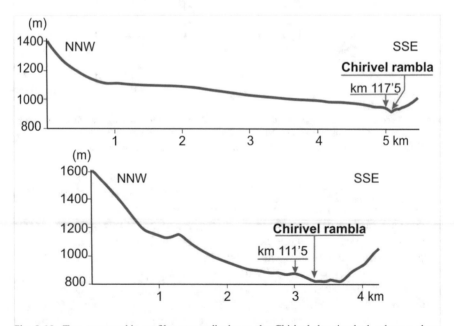

Fig. 2.12 Two topographic profiles perpendicular to the Chirivel dry riverbed, where a sharp increase in the gradient is observed eastwards

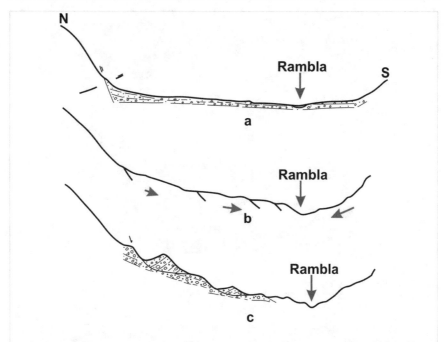

Fig. 2.13 Diagrammatic profiles showing the genesis of the forms; **a** initial situation; **b** erosion of the dry riverbed and the start of instability on slopes; **c** gravitational landslides forming the dolines

deriving from erosion of the Chirivel riverbed, enhanced by saturation of the underlying sandy clay.

The surface of rupture is curved, which causes the strata, initially tipping towards the south, to end up in the opposite dip, up to 40°. The 'dolines' in the loose materials of the debris have the same origin, although in some cases there seem to be casts of loose edges that have generated staggered dolines. There is no evidence of typical karst processes, although some dolines are partially on Jurassic carbonate.

2.7   Box 4: The Pinoso Karst (Alicante)

The Pinoso gypsum (Fig. 2.14) has developed karst in the Triassic diapir of the same name at the intersection of two large orthogonal fractures in the Tertiary. Recent halokinetic movements can be identified, and this indicates that the processes are still active. The diapir is typically dome-shaped (Fig. 2.15) and 7.5 km^2 in area and 893 m at its maximum height, with a base at 540 m.

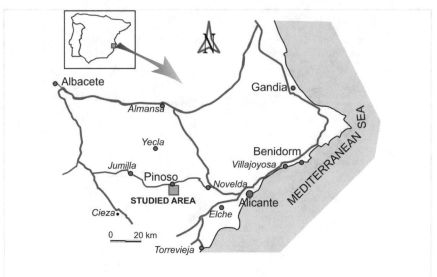

Fig. 2.14 Pinoso diapir

The Compañía Arrendataria de Torrevieja SA has been mining the halite of the Pinoso diapir since the 1980s by conducting leaching in deep boreholes (Fig. 2.16). Water is injected from a nearby aquifer to dissolve the salt (Photo 2.12). The brine that is generated is pumped out, creating a hole. A petroleum derivative is introduced into the borehole to prevent the dissolution from ascending and causing potential risk of collapse. The entire operation is controlled by a sophisticated engineering system, and in 2010, eight boreholes were operational, descending to a depth of about 1200 m.

After physical treatment, the brine is conveyed through a 60 km long pipeline to Torrevieja, where it is evaporated in crystallization ponds. When the dissolution cavity in the borehole becomes very large, a new borehole is drilled. Normally, this switch in location is made when a cavity reaches 80 m in diameter and 600–800 m in height, so its roof can remain 200–300 m from the ground surface The distribution of these workings can contribute to an increased number of dolines; Fig. 2.17 shows in diagrammatic form the type of collapse mentioned above.

Sinkholes are visible in the areas between these boreholes. On the other hand, we can tell which caverns have been induced by borehole exploitation since these take on an inverted cone shape, as shown by the geophysical surveys that are periodically undertaken, in Chap. 3 [1]. At the surface, breakages of sinkholes serve as pathways for rainwater and will accelerate the karstification processes.

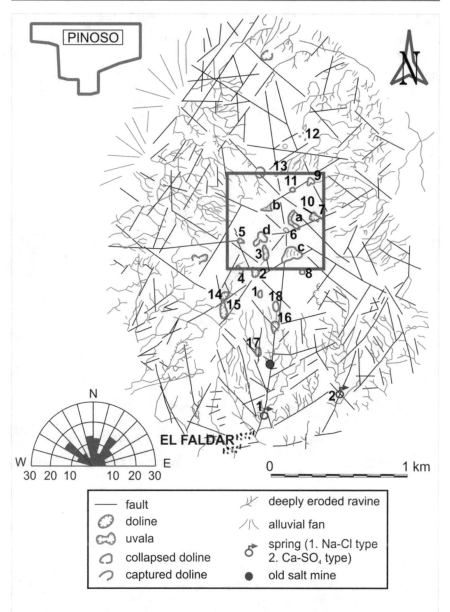

Fig. 2.15 Dome of Pinoso, showing its main fractures and the most relevant exokarst forms. The red box indicates the approximate extent of Fig. 2.16 [23]

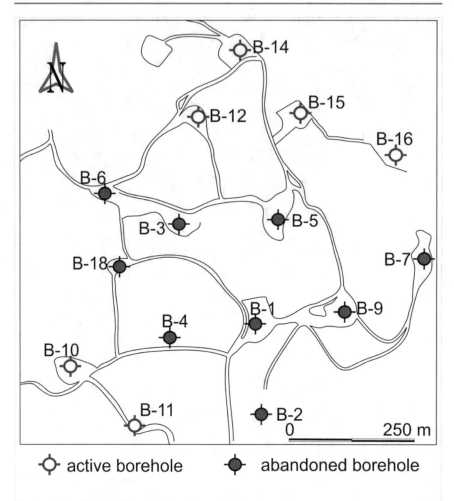

Fig. 2.16 Boreholes at the top of the Pinoso diapir

The quantity of brine extracted is around 600,000 tonnes per year; as the extraction has continued for forty years, since its density is 1.2 t/m³, the volume of the cavity created is around 20 Mm³. This figure coincides approximately with [18] the extent of the cavities that can be deduced from the size of the active and abandoned caverns (17.5 Mm³), based on dimensions of 80 m in diameter by 700 m in height and shaped like an inverted cone.

Photo 2.12 A recent salt exploitation borehole (Photo A. Pulido)

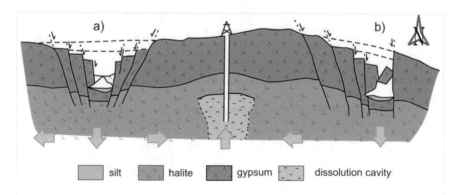

Fig. 2.17 Potential genesis of some visible forms

2.7.1 Geomorphological Aspects

Survey data show that Pinoso gypsum is from 70 to 100 m in thickness, and that it rests on massive halite. There is a radial network of ravines, some of which are deeply embedded, starting at the highest levels and largely following the network of fractures. Maximum erosion is reached in the upper and middle sections, probably favoured by the continuous halokinesis. This ends in a fan that spills down to the foot of the dome. If one compares the directions of ravines and dominant fractures, the structural control of this network is remarkable. Of the 170 measurements taken (Fig. 2.15), those oriented 121–135° and 166–180° (north-south) predominate, coinciding with the diapir's major axes.

Among the karst forms (Photos 2.13, 2.14, 2.15, 2.16, 2.17 and 2.18) that dominate the very variable forms and dimensions, there are 18 dolines more than 5 m in diameter and three main uvalas, some with evidence of cultivation. Two types of sinks can be distinguished in dolines: collapse of the central part, in a series of steps; and lateral collapse, generally in sloping or asymmetric dolines, in this case the highest in the collapsed zone. These dolines reach a depth of 10–20 m.

Most of the dolines and uvalas located near the boreholes are of anthropogenic origin (funnel type); only those that are far from farms or on slopes are natural, and they are shallow [24].

2.7.2 Hydrogeology

The dome shape of the diapir does not favour the development of a karst aquifer [18], in the strict sense, although there are numerous small springs surrounding the diapir at between 600 and 700 m in altitude [25]. The numerous open shapes on the upper parts of the dome favour the rapid infiltration of any run-off generated by rainfall, in a three-phase medium.

Photo 2.13 Characteristic landscape of the upper part of the dome (Photo A. Pulido)

Photo 2.14 Relatively recent doline affecting an long-cultivated bank (Photo A. Pulido)

Photo 2.15 One of the multiple sinkholes on the dome (Photo A. Pulido)

Photo 2.16 Spectacular sinkhole endangering an almond tree (Photo A. Pulido)

Photo 2.17 Detail of the saline deposit left by Spring 2 along its route (Photo A. Pulido)

Photo 2.18 Trail of salt left by a minor escape from a saleoduct (Photo A. Pulido)

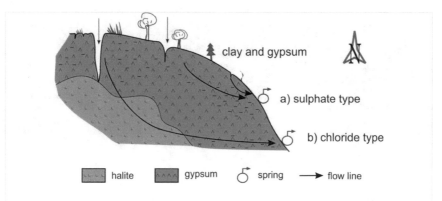

Fig. 2.18 Scheme of the genesis of calcium sulphate-type and sodium chloride springs

Very possibly, the saturated zone is limited to the lowest topographical fraction. The upper fringe is made up of gypsum with a clay fraction that rests on the halite. Groundwater can follow two paths (Fig. 2.18).

1. After infiltrating the dolines and superficial discontinuities, an essentially vertical flow can take place, following a clay-like level. It would be calcium sulphate-type water.
2. Water that penetrates deeper, favouring the discontinuities and tension faults that reach the halite. Following saline impermeable beds, this water

Photo 2.19 Emergence of brackish water from the southern edge, No. 2 of Fig. 2.16 (Photo A. Pulido)

circulates, emerging on a clay bed after the superficial extent of the gypsum has run out. The water would be sodium chloride-type water, an authentic brine. This is the case of the mainspring of the area (No. 2 of Fig. 2.15 and Photo 2.19), for centuries used as a resource for the salt flats established there.

2.8 Further Reading

Appelo, C.A.J., Postma, D. 2005. *Geochemistry, Groundwater and Pollution*, 3th edn. Rotterdam: Balkema, 536 p.

Bögli, A. 1980. *Karst Hydrology and Physical Speleology*. Berlin: Springer-Verlag.

Calaforra, J.M., Pulido-Bosch, A. 1999. Genesis and evolution of gypsum tumuli. *Earth Surface Processes and Landforms*, 24: 919–930.

Cvijic, J. 1893. Das Karstphaenomen. Versuch einer geographischen Morphologie. *Geogr. Abhandlungen von A. Penck*, Vienna, V (3): 217–230.

Dreybrodt, W. 1988. *Processes in Karst Systems*. Berlin: Springer–Verlag, 288 p.

Dreybrodt, W., 1990. The role of dissolution kinetics in the development of karstification in limestone: A model simulation of karst evolution. *Journal of Geology*, 98: 639–655.

Gabrovsek, F. 2007. On denudation rates in Karst. *Acta Carsologica*, 36: 7–13.

Goldscheider, N., Mádl-Szőnyi, J., Erőss, A., Schill, E. (2010) Review: Thermal water resources in carbonate rock aquifers. *Hydrogeology Journal,* 18(6): 1303–1318.

Klimchouk, A. 2015. The karst paradigme: Changes, trends and perspectives. *Acta Carsologica,* 44 (3): 289–313.

Hill, C.A., Forti, P. 1997. *Cave Minerals of the World,* 2nd edn. Huntsville, USA: National Speleological Society, 463 pp.

Wigley, T.M.L. 1973. The incongruent solution of dolomite. *Geochim. Cosmochim. Acta,* 37, 1397–1402.

White, W.B. 2007. Evolution and age relations of karst landscapes. *Acta Carsologica,* 36, 45–52.

2.9 Short Questions

1. List the major groups of processes involved in karstification.
2. Why does water with a high level of organic matter has increased karstification power?
3. What influence does temperature have on karstification?
4. Are limestones and dolomites equally karstifiable? Explain your answer.
5. Why is it not easy to find endokarst formed of large accumulations of common salt?
6. What role do insoluble substances play in karst morphology?
7. List the main exokarst forms that you know.
8. Cite the endokarst forms that you know. Differentiate between the forms of erosion-corrosion and reconstruction.
9. Which endokarst forms are of most hydrogeological importance?
10. What role does erosion play in karstification?
11. What are gypsum tumuli and how do they form?
12. Why are carbonate concretions often found in gypsum ducts?
13. Why can rooted diapirs display a very complex hydrogeological and hydrogeochemical behaviour?
14. Explain the possible origins of poljes.
15. What is a ponor, and what hydrodynamic role does it play?

2.10 Personal Work

1. Karst denudation and the long-term evolution of a karst massif.
2. Karst, and its associated dangers and risks: examples from around the world.
3. Unusual examples of karst in gypsum.
4. Examples of karst in hypersoluble rocks around the world.

References

1. Appelo, C. A. J., & Postma, D. (1993). *Geochemistry, groundwater and pollution* (2nd ed. 1999; 3nd 2005, p. 536). Rotterdam: Balkema.
2. Dreybrodt, W. (1988). *Processes in karst systems* (p. 288). Berlín: Springer-Verlag.
3. Bögli, A. (1980). *Karst hydrology and physical speleology.* Berlin: Springer.
4. Fagundo, J. R., Valdés J. J., & Rodríguez, J. E. (1996). *Hidroquímica del Karst* (p. 304). University of Granada.
5. Calaforra, J. M., & Pulido-Bosch, A. (1996). Some examples of gypsum karsts and the more important gypsum caves in Spain. *International Journal of Speleology, 25*(3–4), 225–237.
6. Dronkert, H. (1976). Late Miocene djointes in the Sorbas basin and djoin áreas. *Bollettino della Società Geologica Italiana, 16,* 341–362.
7. Calaforra, J. M., & Pulido-Bosch, A. (1997). Peculiar landforms in the gypsum karst of Sorbas (Southearn Spain). *Carbonates and Evaporites, 12*(1), 110–116.
8. Calaforra, J. M., & Pulido-Bosch, A. (1999). Genesis and evolution of gypsum tumuli. *Earth Surface Processes and Landforms, 24,* 919–930.
9. Pulido-Bosch, A. (1986). Le karst dans les gypses de Sorbas (Almería). Aspects morphologiques et hydrogéologiques. *Karstologie, Memoires, 1,* 27–35.
10. Calaforra, J. M. (1998). *Karstología de yesos* (Vol. 3, 384 p). Universidad de Almería— Instituto de Estudios Almerienses eds. Serie Monografías Ciencia y Tecnología.
11. Pulido-Bosch, A., Calaforra, J. M., Pulido-Leboeuf, P., & Torres-García, S. (2004). Impact of quarrying gypsum in a semidesert karstic area (Sorbas, SE Spain). *Environmental Geology, 46* (5), 583–590.
12. Calaforra, J. M., & Pulido-Bosch, A. (2000). Cave development in vadose settings in a multilayer aquifer—The Sorbas karst, Almería, Spain. In A. B. Klimchouk, D. C. Ford, A. N. Palmer, & W. Deybrodt, (Eds.), *Speleogenesis Evolution of Karst Aquifers* (pp. 382–386). National Speleological Society.
13. Calaforra, J. M. (1995). El Sistema Covadura (Karst en yesos de Sorbas). *Tecnoambiente, 48,* 73–80.
14. Calaforra, J. M. (2003). *El karst en yeso de Sorbas, un recorrido subterráneo por el interior del yeso* (p. 83). Consejería de Medio Ambiente de la Junta de Andalucía.
15. Pulido-Bosch, A. (1979). *Contribución al conocimiento de la hidrogeología del Prebético Nororiental (provincias de Valencia y Alicante)* (Vol. 95, p. 410). Servicio Publicaciones Ministerio de Industria. Memorias IGME. ISBN 84–7474–050–9.
16. Pulido-Bosch, A. (1977). El karst en yesos de Vallada (Valencia). Incidencia en la calidad química de las aguas. *Cuadernos de Geología, 8,* 113–124.
17. Carulla, N. (1977). *Contribución al conocimiento de la dinámica hidrogeológica en clima semiárido (Depresión de Vera, Almería)* (p. 373). Tesis Doct., Universitat Autònoma de Barcelona.
18. Donat, J. (1966). Río subterráneo "Túnel del Sumidor" (Vallada, Valencia). *Arch. Prehist. Levant., XI,* 255–273.
19. Vilanova, J. (1893). *Memoria Geognótica-Agrícola y Protohistórica de Valencia* (p. 488). Madrid: Sociedad Geográfica.
20. Moreno, I. (1980). *Contribución al conocimiento hidrogeológico de las sierras de María y del Maimón (provincia de Almería)* (p. 194). Tesis Lic., University of Granada.
21. Moreno, I., Pulido-Bosch, A., & Fernández Rubio, R. (1983). Hidrogeología de las sierras de María y del Maimón (provincia de Almería). *Boletín Geológico Minero., XCIV*(IV), 321–338.
22. Moreno, I., & Pulido-Bosch, A. (1982). Formas "exokársticas" en materiales de piedemonte del borde meridional de la sierra del Maimón (Vélez–Rubio, Almería). *Reunión Monográfica del Karst de Larra, II,* 129–138.
23. Rodríguez Estrella, T. (1983). Neotectónica relacionada con las estructuras diapíricas en el Sureste de la Península Ibérica. *Tecniterrae, 51,* 14–30.

24. Navarro Hervás, F., & Rodríguez Estrella, T. (1985). Características morfoestructurales de los diapiros triásicos de Hellín, Ontur, la Celia, Jumilla, La Rosa y Pinoso, en las provincias de Albacete, Murcia y Alicante. *Papeles de Geografía, 10*, 49–56.
25. Rodríguez-Estrella, T., & Pulido-Bosch, A. (2010). Gypsum karst evolution in a diapir: a case study. (Pinoso, Alicante, Spain). *Environmental Earth Sciences, 59,* 1057–1063. https://doi.org/10.1007/s12665-009-0097-2.

Conceptual Models of Karst Aquifers

3

3.1 Glossary

Martel's model considers that there are underground rivers only in limestone massifs.

Sokolov's model distinguishes four circulation zones in karst aquifers: aeration, where perched karst water can be found; seasonal fluctuation; complete saturation; and deep circulation.

Mangin's model recognizes a **hierarchical network** in a **karst system** in the form of an underground river and lateral feeding sources, which he terms **annexed systems** that may or may not be karst terrain.

The block model is of large **blocks** of low permeability, separated by discontinuities of high permeability (**conduits**), assuming a low volumetric percentage. The largest volumes of water stored in karst systems are in the blocks, corresponding to the **capacitive element**, while the **transmissive element** is provided by the network of conduits.

An **intermediate model** combines the block and conduit model to the hierarchical model, proposing that there is a higher k-value in the vicinity of discharge areas than in recharge areas.

There is a marked effect of **scale** on the permeability of karst aquifers, with the highest values being seen at the basin scale.

The **specific flow** of a borehole is the quotient of the pumping flow and the decrease in flow that it generates (l/s/m). The specific unit flow is the specific flow divided by the thickness collected (l/s/m/m).

© Springer Nature Switzerland AG 2021
A. Pulido-Bosch, *Principles of Karst Hydrogeology*, Springer Textbooks in Earth Sciences, Geography and Environment,
https://doi.org/10.1007/978-3-030-55370-8_3

3.2 General Aspects

We have already seen that discontinuities play a foundational role in karstification. The main discontinuities of hydrogeological interest are fractures, fissures and stratification surfaces. These influence the extent of infiltration, transit, storage—together with the effective porosity of the rock—and circulation in the saturated karst medium. Some of the first observers of the subterranean environment wondered if there are aquifers in the karst or if there are only *subterranean rivers*. This is what is referred to as the *Martel model*, named after the great speleologist of the nineteenth and twentieth centuries. In general, it was the early speleologists who, due to the partial aspect of their observations, described the absence of hydrogeological continuity in the karst medium, admitting only a few networks of conduits with hardly any relation between them in the form of river basins. Among the first proponents of a general saturation level, or aquifers, in the karst were Cvijic [1, 2] and Grund [3].

Today, it is generally accepted that there is a saturation zone in the karst. Thus, any water from rain or another source that reaches the surface will infiltrate those discontinuities (sinkholes, sinks, open fractures and so on) with an essentially vertical route through the *unsaturated zone*, until it reaches the *saturated zone*. In the saturated zone, its movement is governed by the hydraulic head at any point, and the flow is from those areas with the greatest to those with the least potential, from the recharge zones to the discharge zones, in a predominantly horizontal movement.

Sokolov [4], on the other hand, recognizes four zones: of *aeration*, where karst waters can be perched; of *seasonal fluctuation;* of *complete saturation*; and of *deep circulation*. The last has few known characteristics. It corresponds to regional circulation in Toth's hydrodynamic scheme [5]. Figure 3.1 shows the most widespread conception of k variation with depth, largely supported by civil engineering pumping test results in Chap. 1 [14].

Thus, in the aeration zone, vertically developed discontinuities and the stratification planes play an important role in the transmission of water, and to the decompression and alteration zone if there is an epikarst aquifer. In the saturated zone, these are horizontally developed discontinuities. In all cases, fracturing plays a foundational role.

3.3 Conceptual Models

3.3.1 General Characteristics

Mangin in Chap. 1 [18] regards the karst system as a *hierarchical* network, as an underground river yet one that has a piezometric level and highly diverse sources of lateral feeding, which he terms *annexed systems,* which may or may not be karst. An example would be a river on granite, which loses its waters down a swallow

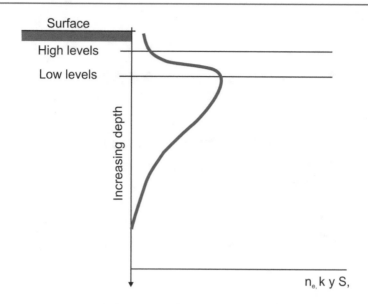

Fig. 3.1 Variation in Ne, K and S at depth (Adapted from in Chap. 1 [14])

hole in contact with limestone (Fig. 3.2). In the latter case, we have a *binary* system. Both Mangin and Bakalowicz [6, 7] consider that the karstification zone is essentially restricted to the piezometric fluctuation fringe, unless there are actual or have been past changes in base level. The system would cease to be 'functional' when the network becomes enlarged sufficiently to evacuate the water that it receives as recharge.

A more appropriate conceptual model for a karst aquifer is to consider it as made up of very large blocks of low permeability, separated by a network of highly permeable discontinuities yet which involves a low volumetric percentage. The large volumes of water in karst systems are essentially be stored in the blocks, as the *capacitive element*, while the *transmissive element* would be provided by the conduit network [8]. This would explain why, in a karst environment, there are high-yield boreholes alongside others that are very low yield.

From this first simple model, it is possible to conjecture about others, including the *intermediate*-type that combines the block and conduit model with the *hierarchical model*, resulting in a higher *k-value* in the vicinity of discharge areas compared to recharge areas (Fig. 3.3). Figure 3.4 shows the model proposed by Kiraly [9–11], which is a combination of all models, relatively less concerned with the aquifer's geometry.

This leads us to consider the question of scale in the karst medium, and the limits of validity of the determinants of the physical parameters in each. Thus, for example, determining *k* in laboratory samples will give us the *k* that is normally referred to in blocks; on a borehole scale, we will have the entire range between

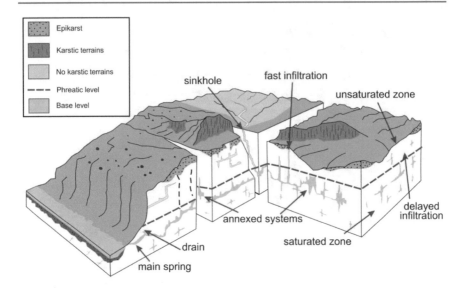

Fig. 3.2 Model of the karst system and its relationship to annexed systems (Adapted from in Chap. 1 [18])

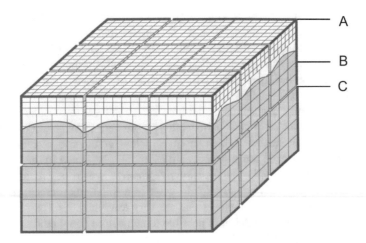

Fig. 3.3 Conceptual block and conduit model: **a** Epikarst; **b** medium fractures within the block; **c** large fracture or conduit (Adapted from [9])

large conduits and blocks; and finally, on a basin scale, we will have an integration of the whole (Fig. 3.5).

The values of the storage coefficient vary greatly and are strongly dependent on the lithology, climate, structure and so on. The most common values are between 0.5

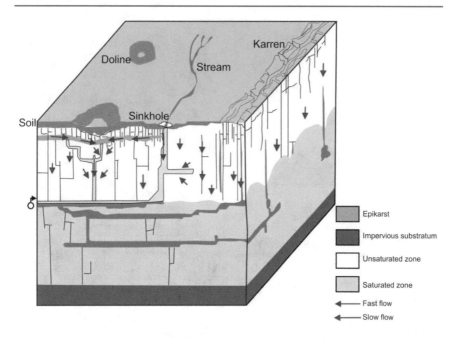

Fig. 3.4 Schematic diagram of a fissure karst aquifer (After [10])

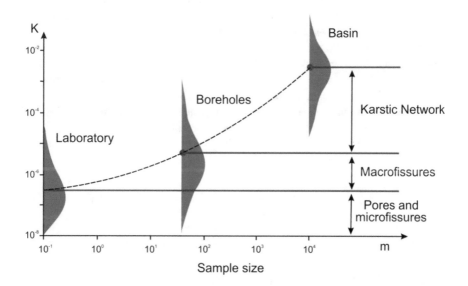

Fig. 3.5 Effect of scale on the permeability of karst aquifers (Adapted from [9])

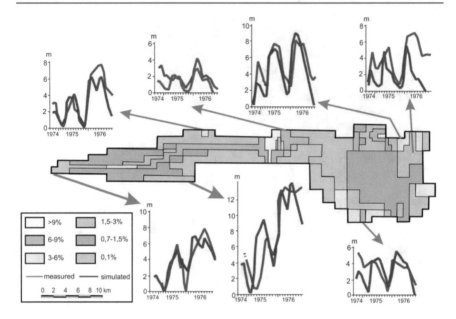

Fig. 3.6 Sierra Grossa karst aquifer simulation model (Valencia [12])

and 1%. Very exceptionally they exceed 5%, always in highly transmissive sections or in very porous dolomite. Figure 3.6 shows a simulation of the Sierra Grossa aquifer that uses an *intermediate*-type conceptual model, recognizing the hierarchy of karstification from recharge areas (less karstification) to discharge areas (maximum karstification). Both transmissivity T and storage coefficient S are regarded as increasing in the discharge-recharge direction, but it is conceptually necessary to resort to highly transmissive elements, represented by two orthogonal fringes.

To illustrate the process of arriving at a hypothesis using the conceptual model of a karst aquifer, a synthesis of the work by Pulido-Bosch and Castillo [13] is included below, based on data from boreholes drilled in karst aquifers in eastern Spain (provinces of Valencia and Alicante).

3.4 Box 1: Data to Build a Conceptual Model

3.4.1 Characteristics of Boreholes

The carbonate aquifers (limestone and dolomite) of the Spanish Levant are aquifers of great economic interest. Essentially, they are tapped by both percussion and rotary percussion boreholes, and 143 of these boreholes in the

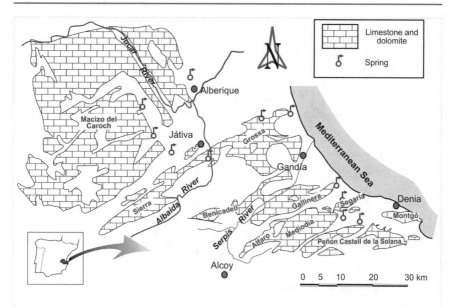

Fig. 3.7 Schematic map of the hydrogeological units mentioned: (1) Outcrops of the Creu formation; (2) Springs of a flow of greater than 100 l/s

area in Fig. 3.7 were the object of a detailed statistical study. They have in common the fact that they capture water from the Creu formation: dolomite—about 300 m thick—calcarenite, sandstone and sandy marl—70 to 120 m thick—with a white limestone at the bottom, 120 m thick in Chap. 2 [15].

The same figure shows a diagrammatic distribution of the Creu formation outcrops in this studied sector. They constitute five hydrogeological units: Caroch Massif (1800 km^2), Sierra Grossa (370 km^2), Benicadell-Almirante-Gallinera (230 km^2), Almirante-Mediodía-Segaria (175 km^2) and Peñón--Castell de la Solana-Montgó (100 km^2).

The materials of the Creu formation have been folded and fractured; the units listed form anticlines, separated by synclines of variable size. The materials constitute a substantial set of white and blue marls ('tap') from the Miocene (Belgida formation), which keep the units separate. In these, units are springs (Fig. 3.8) of very different flow rates and recharge area extent.

Of the total of 143, 113 boreholes and wells were examined. We know their drilled depth, their saturated depth drilled in limestone or dolomite, and their yield. In addition, in 85 cases we know the drawdown for the pumping flow rate, which allows the specific yield of the borehole to be calculated. The hydrogeological reports of 42 boreholes were available, and data on the remaining 71 were obtained from the owners.

For some calculations, boreholes in the discharge area were treated differently from those in recharge areas. This decision was made arbitrarily on the basis of parameter $L/10$, the radius centred on the spring, where L is the

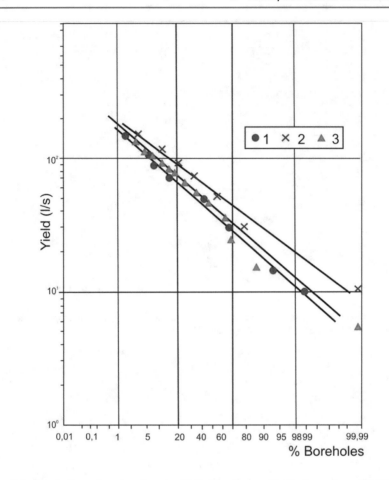

Fig. 3.8 Flow rates on log-normal paper: (1) Recharge zone; (2) Discharge zone; (3) All data

maximum length of the hydrogeological unit (Table 3.1). Only the carbonate rock **thickness** has been considered, discounting the marl levels in the saturated zone and the total unsaturated zone.

The maximum thickness of the carbonate is 360 m and the minimum 2 m; the average is 108.5 m; and the standard deviation is 71.9 m. By unit, the greatest thickness is in the Caroch Massif (121.6 m), followed by Sierra Grossa (115.4 m); the least is that of Benicadell-Almirante-Gallinera (85.9 m). The average in discharge areas is 103.3 m ($S = 74.0$), while in the recharge area it is 113.7 m ($S = 69.7$). In order to compare and differentiate the thicknesses recorded on the basis of whether it is a recharge or discharge area and also by hydrogeological unit, a test was carried out on the averages, taking as a null hypothesis the equality of the averages, for a significance level of 5%.

Table 3.1   Distribution of boreholes and wells by units and zone

Hydrogeological unit	Borehole		Total
	Recharge	Discharge	
Caroch Massif	14	9	23
Sierra Grossa	22	25	47
Benicadell-Almirante-Gallinera	11	13	24
Alfaro-Midday-Segaria	0	9	9
Penón-Castell de la Solana-Montgó	9	1	10

The *flow rates* of the boreholes are between 250 and 3 l/s, with an average value of 46.7 l/s and a standard deviation of 37.9 l/s. It is noted that 4.53% of the values do not exceed 40 l/s, and only 8.9% exceed 100 l/s. The average flow rate in the discharge area is 52.6 l/s ($S = 36.1$), while in the recharge area, it is 40.7 l/s ($S = 39.0$). However, according to the test performed for a significance level of 0.05, the difference is not significant.

The flow distributions as a whole and by area have been adjusted to a log-normal distribution by means of the X^2 test. For all data, the parameters are 3.49 and $\sigma = 0.05$; for the discharge zone: $= 3.73$ and $\sigma = 0.85$; for the recharge zone: 3.26 and $\sigma = 0.99$. In all cases, the log-normal distribution of flows for a significance level of 0.01 can be accepted (Fig. 3.8).

The flow rate ratio of the thickness captured, decimal or logarithmic (Figs. 3.9 and 3.10), is clearly scattered, as shown in Fig. 3.11. The range of most frequent values is 40–60 l/s and 50 m of thickness.

In 85 of the 113 boreholes examined, the drawdown for the pumping yield was known; the specific yield values obtained are highly variable, ranging between 0.048 and 128 l/s/m; the mean value is 11.15 l/s/m; and the standard deviation is 19.7.

It is noted that more than 65% of the boreholes have a **specific yield** of less than 10 l/s/m, and only slightly more than 7% exceed 20 l/s/m. The highest average values per unit correspond to Caroch (23.8 l/s/m), and the lowest to Castell de la Solana-Peñón-Montgó (5.8 l/s/m), with the largest and smallest total area, respectively. The values are clearly different in the recharge zone ($n = 43$; mean 4.9 l/s/m; $S = 5.0$) from the discharge zone (Fig. 3.9; $n = 42$, mean 17.4 l/s/m and $S = 26.1$). They are statistically different averages, with a significance level of 0.05. This pattern is repeated in each hydrogeological unit (Table 3.2). From the morphology of the histograms, we can deduce that its distribution is not log-normal; it is multimodal without positive asymmetry, as confirmed analytically.

As with the flows, we have tried to determine the specific yield–thickness ratio. In no case does the linear correlation coefficient attain a significant value: -0.042 for all data; -0.009 for the discharge zone; and -0.134 for recharge zone. Similar results are obtained by correlating the thickness and Naperian logarithm of the specific yield Q_s ($r = 0.252$) or the Naperian

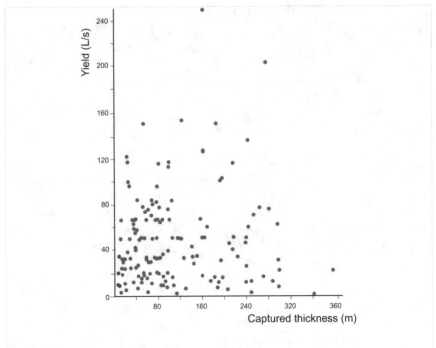

Fig. 3.9 Flow rates as a function of the captured thickness

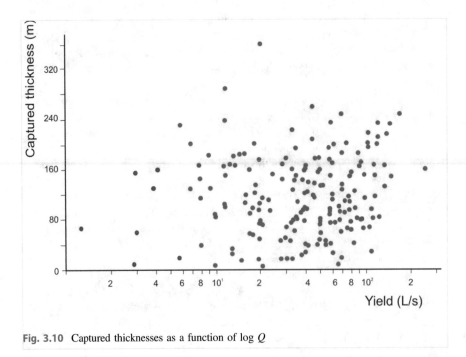

Fig. 3.10 Captured thicknesses as a function of log Q

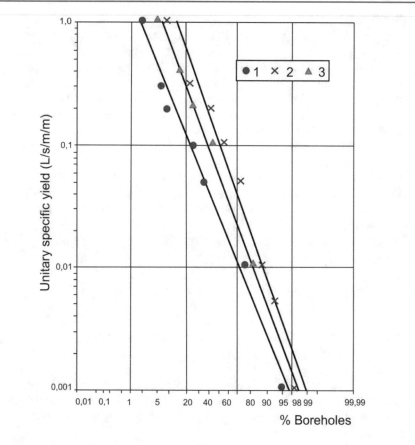

Fig. 3.11 Unitary specific yields in log-normal probabilistic paper. (1) Recharge zone; (2) Discharge zone; (3) All data

Table 3.2 Means and standard deviations of specific yield rates by units and zones

Hydrogeological unit	Discharge		Recharge	
	μ	S	μ	S
Caroch Massif	52.1	53.1	4.3	4.7
Sierra Grossa	8.4	6.6	7.3	5.9
Benicadell-Almirante-Gallinera	8.6	4.7	2.3	1.1
Alfaro-Mediodía-Segaria	9.1	5.6	–	–
Penón-Castell de la Solana-Montgó	25.0	0	3.4	3.7

logarithm of the thickness with that of Q_s ($r = 0.167$), or even by taking the flow and logarithm of the recorded thickness ($r = 0.0129$).

We also analysed the **specific 'unitary' yield**, the result of dividing the specific yield by the metres drilled into the carbonate aquifer. The average of the 85 instances is 0.239, with a standard deviation of 0.506. The minimum value is 0.0003 l/s/m and the maximum 2.84 l/s/m. In the discharge zone, the average is 0.4104 l/s/m ($S = 0.718$), while in the recharge zone it is 0.09957 ($S = 0.1774$). A test on the difference of the average values concludes that the null hypothesis cannot be accepted ($\mu_1 = \mu_2$), at the level of significance 0.01; there is therefore a statistically significant difference of average values. The specific unitary yield in a homogeneous medium will relate to the permeability of the aquifer. We verified that a log-normal distribution is followed in the entire sample and in the boreholes in both the recharge and in discharge areas, with a significance level of 0.01. The parameters of the adjusted log-normal distributions are: $\eta = 2$–0.04 and $\sigma = 1.95$ (sample total); $\eta = 2$–0.26 and $\sigma = 1.88$ (discharge zone); and $\eta = 3$–0.61 and $\sigma = 1.78$ (recharge zone). Figure 3.9 shows the adjusted lines on a log-normal scale.

3.4.2 Factors Influencing the Structure of Aquifers

From the foregoing, it can be deduced that the values analysed have a wide dispersion. Apparently, any yield can be obtained with any recorded thickness. We also see that the various types of yields analysed have higher values in the discharge area. Finally, we have shown that yields and specific yields follow a log-normal distribution.

In addition, karstification processes act on fractured rock and other discontinuities, contributing to their selective widening [14]. The rock changes from a very heterogeneous and anisotropic medium to one that is less so, even one that is capable of being considered homogeneous at a certain scale. A fissured carbonate aquifer can evolve into a karstified aquifer with a more or less hierarchical network of conduits. Depending on the degree of hierarchization, it results in a karst aquifer or simply a fissured one

The fact that yields and specific yields have a log-normal distribution, as well as the distribution of the cumulative length of fractures, can be interpreted as evidence of the influence of fracturing on karstification and thus on the performance of boreholes. During drilling, if we intercept a fracture of a kilometre in length it leads us to think that its performance will be greater than one that is just tens of metres long.

Furthermore, the evidence that specific yields and unitary specific yields attain higher average values in discharge areas than in recharge areas indicates that karstification, although it acts on the entire massif, is more intense in the former. We interpret this as proof of a hierarchy in the flow and the conduits, although not comparable to a surface drainage network, at least not

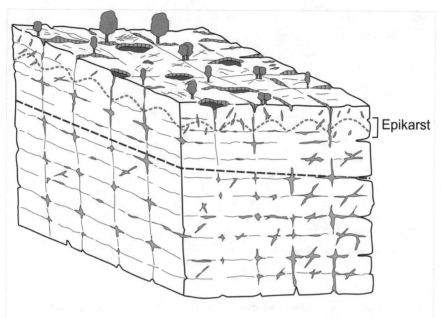

Fig. 3.12 Conceptual model of karst aquifer of blocks and conduits, with hierarchical flow

in the massifs studied here. This allows us to propose a conceptual model of a karst aquifer of blocks and conduits, with hierarchization of the flow from the recharge to the discharge zones, which is intended to be represented in Fig. 3.12.

The various aspects studied are of great economic application to the whole Mediterranean area, especially when planning the sites of new boreholes. If we take into account the relatively low topography of the discharge areas, the boreholes can be less deep and the elevation required lower, so the exploitation costs will be less.

3.5 Box 2: Typology of Aquifers in Carbonate Rock

In the real world, karstification processes are highly dynamic and begin to act during the consolidation and diagenesis of sediments. The factors that influence its evolution are both intrinsic and extrinsic. Among the former are the sedimentation medium: an oolitic limestone is not the same as a micritic limestone. Among the latter, tectonics and climate are possibly the most influential. The fissuring and fracturing that accompany tectonic stresses—as already seen—promote the access of water into the aquifer. Climatic conditions, influencing the rainfall, are also key. This is why karstification is most

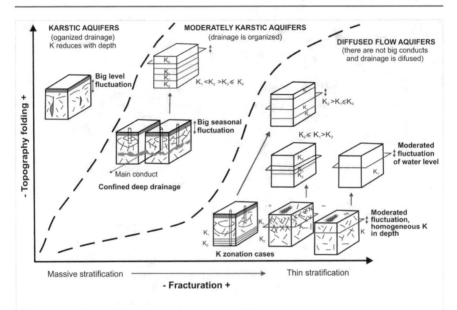

Fig. 3.13 Typology of aquifers in carbonate rocks, as proposed by Bayó [15, 16], showing their hydrodynamic characteristics

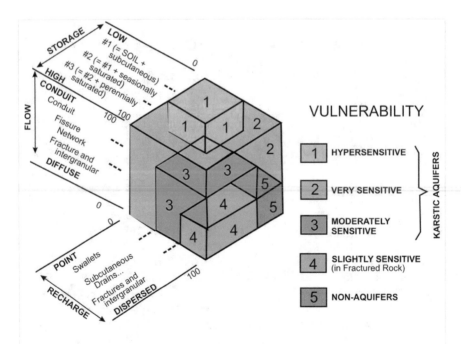

Fig. 3.14 Alternative typology of aquifers in fissured media, on the basis of storage, flow and recharge, with implications for their susceptibility to contamination (Adapted from [17])

active in regions with a humid temperate climate, also promoted by the high level of biological activity. Figures 3.13 and 3.14 illustrate two possible classifications of aquifers in the natural environment; the second includes the concept of vulnerability, which will be studied in Chap. 8.

3.6   Further Reading

Cvijic, J. 1918. Hydrographie souterraine et evolution morphologique du karst. *Recueil des Travaux de l'Institut de Gographie Alpine*, 6: 375–429.

Grund, A. 1903. Die Karsthydrographie. Studien aus Westbosnien. *Geogr. Abhandlungen von A. Penck*, VII, 3: 1–201.

European Commission. 1995. *Karst Groundwater Protection*. COST Action-65. Report EUR 16547 EN, 446 p.

Jeannin, P.Y., Eichenberger, U., Sinreich, M., Vouillamoz, J., Malard, A., Weber, E. 2013. KARSYS: A pragmatic approach to karst hydrogeological system conceptualisation. Assessment of groundwater reserves and resources in Switzerland. *Environmental Earth Sciences*, 69 (3): 999–1013.

Jeelani, G., Kumar, U., Bhat, S.N.A., Sharma, S., Kumar, B. 2015. Variation of δ18O, δD and 3H in karst springs of south Kashmir, western Himalayas (India). *Hydrological Process*, 29 (4): 522–530.

Williams, P.W. 2008. The role of the epikarst in karst and cave hydrogeology: A review. *International Journal of Speleology*, 37: 1–10.

3.7   Short Questions

1. What is Martel's karst model?
2. Describe the vertical zoning of karst aquifers.
3. Do you agree with the presence of a hierarchical network in karst?
4. Do you agree with the existence of annexed systems? Justify your answer.
5. What is the base level, and what are its consequences?
6. Do you agree with the classification into capacitive and transmissive elements in the karst? What consequences does this have on underground works?
7. What is epikarst, and what role does it play in a karst aquifer?
8. What is the conceptual model of blocks and ducts in karst aquifers?
9. Briefly explain the effect of the scale on the permeability of karst aquifers and its applied consequences.
10. What is the specific flow of a borehole?
11. What is the specific unit flow of a borehole?

12. Does it make sense to differentiate between recharge and discharge areas when designing collection boreholes? Give reasons for your answer.
13. Considering the historical evolution of a carbonate massif, can you distinguish between types of aquifers?
14. To what extent does the historical evolution of a karst massif affect contamination of the massif?
15. What role do clays play in the evolution of karst?

3.8 Personal Work

1. The evolution of groundwater in limestone and dolomite: the main milestones.
2. The main conceptual models of a karst aquifer, and their rationales.
3. The secular evolution of karst aquifers in humid regions: the influencing factors.
4. The secular evolution of karst aquifers in semi-arid regions: the influencing factors, with real-world examples.

References

1. Cvijic, J. (1893). Das Karstphaenomen. Versuch einer geographischen Morphologie. *Geogr. Abhandlungen von A. Penck, 3*, 217–230.
2. Cvijic, J. (1918). Hydrographie souterraine et evolution morphologique du karst. *Recueil des Travaux de l'Institut de Gographie Alpine, 6*, 375–429.
3. Grund, A. (1903). Die Karsthydrographie. Studien aus Westbosnien. *Geogr. Abhandlungen von A. Penck, 3*, 1–201. (Leipzig, VII).
4. Sokolov, D. S. (1965). Hydrodynamic zoning of karst water. *AIHS-UNESCO, I*, 204–207 (París).
5. Toth, J. (1963). A theoretical analysis of groundwater flow in a small drainage basins. *Journal of Geophysical Research, 68*(8), 4795–4812.
6. Bakalowicz, M. (1979). *Contribution de la géochimie des eaux à la connaissance de l'aquifère karstique el de la karstification* (p. 269). Paris IV: Thèses Université.
7. Bakalowicz, M. (2005). Karst groundwater: A challenge for new resources. *Hydrogeoly Journal, 13*, 148–160.
8. Drogue, C. (1980). Essai d'identification d'un type de structure de magasins carbonatés fisurés. Application à l'interprétation de certains aspects du fonctionnement hydrogéologique. *Mémoires de la Société géologique de France, II*, 101–108.
9. Kiraly, L. (1975). Rapport sur l'état actuel des connaissances dans le domaine des caractères physiques des roches karstiques. In *Hydrogéologie des terrains karstiques* (pp. 53–67). Paris: AIH.
10. Kiraly, L. (1988). Large scale 3-Dgroundwater flow modeling in highly heterogeneous geologic medium. In E. Custodio, et al. (Eds.), *Groundwater flow and quality modelling* (pp. 761–775). London: D. Reidel Publishing Company.
11. European Commission. (1995). *Karst groundwater protection* (p. 446). COST Action-65. Report EUR 16547 EN.
12. Pulido-Bosch, A. (1989). Simulación del acuífero de Sierra Grossa (Valencia). *Hidrogeología y Recursos Hidráulicos, XIV*, 301–313.

13. Pulido-Bosch, A., & Castillo, E. (1984). Quelques considérations sur la structure des aquifères carbonatés du levant espagnol, d'après les données de captages d'eau. *Karstologia, IV*, 38–44.

14. Grillot, J. C., & Drogue, C. (1997). Sur le rôle de la fracturation dans l'organisation de certains phénomènes karstiques souterrains. *Karstologie*, 11–22.

15. Bayó, A. (1982). La exploración hidrogeológica de acuíferos en rocas carbonatadas desde la óptica de la exploración de recursos y utilización del almacenamiento. *Reunión Monográfica Karst Larra*, 177–215.

16. Bayó, A., Castiella, J., Custodio, E., Niñerola, S., & Virgós, L. (1986). Ensayo sobre las diversas tipologías de acuíferos en rocas carbonatadas de España. Identificación, técnicas de estudio y formas de captación y explotación. *Jornadas sobre el Karst en Euskadi, 2*, 255–340.

17. Quinlan, J. F. Q., Smart, P. L., Schindel, G. M., Alezander, C. E., Edwards, A. J., & Smith, A. R. (1991). Recommended administrative/regulatory definition of karst aquifers, principles for classification opf carbonate aquifers, practical evolution of vulnerability of karst aquifers, and determination of optimum sampling frequency at springs. In *Proceedings of the 3rd Conference on Hydrology, Ecology, Monitoring and Management Ground Water in Karst Terranes* (pp. 573–635). Dublin, Ohio: National Ground Water Association.

Analysis of Hydrographs

4

4.1 Glossary

The *hydrograph* of a spring is normally regarded as $Q = f(t)$. After rain, there is an increase in the upwelling flow, reaching a maximum (*peak*), then *decreasing* and subsequent *depletion*, or the period when discharge is uninfluenced by rainfall. The whole constitutes the *emptying curve*.

The period after the peak is called the *recession curve (decreased + depletion)*.

The integration of the line obtained between the beginning of depletion and infinity lets us calculate the *volume of water drained by gravity.*

In a **karst system,** a distinction is drawn between the *infiltration subsystem*, corresponding to the unsaturated zone, in which variations in flow (*q*) are responsible a spring's floods and decreases; and the *saturated subsystem*, corresponding to the saturated zone, whose outflow through a spring is Q.

Regulatory power (*k*) is the capacity of the system to progressively restore part of the volume of water passing through it (**transit volume**).

The *index i* characterizes a system by the shape of its mean decreasing curve.

4.2 Introduction

4.2.1 General Aspects

For a long time, studies of karst hydrogeology were based essentially on analysis of a spring's hydrographs. This analysis was the basis for elaborating on the various hypotheses on the structure and function of the karst environment. Studies were undertaken on the spring's depletion (the discharge period uninfluenced by precipitation [1–3]); on its recession or decreasing curve (the period following the peak

© Springer Nature Switzerland AG 2021
A. Pulido-Bosch, *Principles of Karst Hydrogeology*, Springer Textbooks in Earth Sciences, Geography and Environment,
https://doi.org/10.1007/978-3-030-55370-8_4

Table 4.1 Main hydrography study methods

Type	AUTHOR	FORMULA	OBSERVATIONS
BASEFLOW	MAILLET	$Q_t = Q_0 e^{-\alpha t}$	linear representation of log Q vs t; the slope of the line is 0,4343 α. $$V(m^3) = \frac{Q_s\,(m^3/s)}{\alpha(dias^{-1})} \times 86.400$$
BASEFLOW	TISON, WERNER AND SUNDQUIST	$Q_t = \dfrac{Q_0}{(1+\alpha t)^2}$	linear representation of $1/\sqrt{Q_t}$ vs t
DEPLETION	SCHOELLER	$Q_t = Q_{01} e^{-\alpha_1 t_1} + ...$ $... + Q_{0n} e^{-\alpha_n t_n}$	linear representation by sectors; $V(m^3) = \sum_{i=1}^{n} \dfrac{Q_{0i}}{\alpha_i} \times 86.400$ $\alpha = \beta Q_0^2$
DEPLETION	FORKASIEWICZ AND PALOC	$Q_t = \dfrac{1}{\left(\dfrac{1}{Q_0^2 \beta t}\right)^{1/2}}$	linear representation $1/Q_0^2$ vs t; $V_0 - V_t = \dfrac{2}{\beta}\left(\dfrac{1}{Q_t} - \dfrac{1}{Q_0}\right) \times 86.400$
DEPLETION	DROGUE	$Q_t = \dfrac{Q_0}{(1+\alpha t)^n}$	n takes the values : 1/2, 3/2 and 2
DEPLETION	MANGIN	$Q_t = \varphi_t + \psi_t$ $\varphi_t = Q_0 e^{-\alpha t}$ $\psi_t = Q_0 \dfrac{1-\eta'}{1+\varepsilon t}$	ε = heterogeneity coefficient = $\dfrac{q_0 - q^*}{q_t} - \dfrac{\eta' q_0}{q^*}$ $\eta' = 1/$emptying time
GLOBAL	GALABOV	$Q = \dfrac{a\,t^b}{e^{ct}}$	$b = \left(\dfrac{t_m}{t_c - t_m}\right)^2 ; \dfrac{t_m}{(1-t_m)^2}$ $a = Q_{max}\left(\dfrac{2,72}{t_m}\right)^b$

of the hydrograph [4–6] and (in Chap. 1 [18]); and on the global hydrograph [7]. Table 4.1 summarizes the various methods.

In the same line of study, an analysis of the global hydrograph is the use of correlation and spectral analysis, both simple and crossed [8, 9], a basic element when considering the operation of the system and choosing methodologies, especially to simulate underground flow. The starting elements, in this type of analysis, are daily rainfall data and daily flows of the mainsprings in the system in question. It is necessary to have data on the actual recharge in the system or the effective rainfall, which are difficult to establish precisely. Weekly or monthly records provide some information but, due to the rapid variation in discharge values from the karst medium, major mistakes can be made.

4.2.2 Depletion or Base Flow

After a period of rainfall, the hydrograph of a spring shows an increase in discharge (flooding), reaching a maximum (peak), followed by a decrease in flow (decreasing) and subsequent depletion. The process constitutes the *depletion curve* (Fig. 4.1). The period immediately after the peak is often referred to as the *recession curve* (decreased + depletion).

Maillet [1] likens the depletion of an aquifer by base flow to a container of water being emptied through a porous stopper. The initial flow rate Q depends on the

Fig. 4.1 Hydrograph of a spring and its terminology

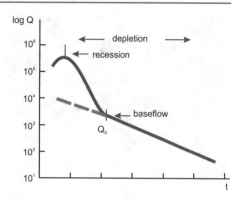

Fig. 4.2 Starting scheme for the deduction of Maillet equation

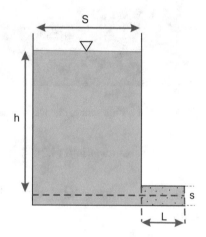

permeability of the stopper k, its section s, the porosity of the material m, the hydraulic head h and, inversely, the length L of the stopper (Fig. 4.2):

$$Q = km\frac{s}{L}h \tag{4.1}$$

If, over a time dt, the level decreases dh, it is equivalent to: $Sdh = -Qdt$ (4.1), where S is the section through the vessel:

$$dQ = km\frac{S}{L}dh \tag{4.2}$$

At the same time, the flow varies if there is a decrease in the hydraulic head. From Eq. (4.1): $dh = \frac{-Qdt}{S}$,

Photo 4.1 La Villa spring, the main drainage in Torcal

which substituted into (4.2):

$$dQ = \frac{-kms}{SL}Qdt; \text{ calling to } \frac{kms}{SL} = \alpha_{us}$$

is:

$$dQ = -\alpha Qdt$$

similar to

$$\frac{dQ}{Q} = -\alpha dt \text{ that integrating between } Q_0, t_0, Q_t$$

and t is obtained:

$$l_nQ_t - l_nQ_0 = -\alpha(t - t_0); \quad l_n\frac{Q_t}{Q_0} = -\alpha(t - t_0)$$

if we take

$$t_0 = 0, l_n = \frac{Q_t}{Q_0} = -\alpha t \text{ equivalent to } Q_t = Q_0e^{-\alpha t} \qquad (4.3)$$

which indicates that the flow decreases exponentially.

Photo 4.2 Gauging station at La Villa spring. This measures the non-derived water fraction for the urban water supply (*Photo* A. Pulido)

The equation lets us ascertain the flow rate at a known time, t, the depletion coefficient, the flow rate at the beginning of the depletion Q_0. Using logarithms in the expression (4.3):

$$\log Q_t = \log Q_0 - \alpha t(\log e); \quad \log Q_t = \log Q_0 - 0.4343\alpha t$$

equation of type $y = ax + b$, from which it can be deduced that by representing log Q as a function of t we will give a straight line (in depletion) of slope 0.4343α. α is usually expressed in days^{-1}.

The integration of the straight line obtained between the beginning of depletion and infinity allows us to obtain *the volume of water that can be drained by gravity*, which should not be confused with the water reserves of the aquifer:

$$V = \int_{t_0}^{\infty} Q_t dt = \int_{t_0}^{\infty} Q_0 e^{-\alpha t} dt = \frac{Q_0}{\alpha}$$

If Q_0 is expressed in m³/s and α in days⁻¹, the volume is:

$$V(m^3/s) = \frac{Q_0(m^3/s)}{\alpha(dias^{-1})} \cdot 86.400$$

4.2.3 Recession

Although there are many possible methods, Schoëller's [4] was one of the first to develop an identification proposal. This author noted that in many cases when log Q is represented as a function of t, a straight section is obtained and also another, which is not. According to the conceptual model of karst, which regards it as a network of conduits and a network of fissures, it is possible to differentiate the flow as due to either one or the other. Therefore, the curvilinear section can be decomposed into as many straight sections of depletion coefficients are appropriate to the size of the conduits (large, medium, small, fissures and so on). In this way, the expression of the void would be a sum of exponentials, each corresponding to a type of conduit:

$$Q_t = Q_{0_1}e^{-\alpha_1 t_1} + Q_{0_2}e^{-\alpha_2 t_2} + Q_{0_3}e^{-\alpha_3 t_3} + \cdots \tag{4.4}$$

This formulation is not very consistent, since there is an interrelation between the various conduits [10]. Drained partial volumes are also deductible for partial integrations.

4.3 Identification of Flow Modalities

4.3.1 Background

The hydrodynamic operation of karst aquifers can be analysed and compared by using Mangin's [9] and Coutagne's [11] equations. These expressions allow us to estimate the volume of water drained over any time interval, as we will see later. We will apply this methodology to four karst springs that drain their respective karst mass at a practically unique discharge point. They are well-defined massifs (in Chap. 2 [15]), [12, 13]. Their characteristics are summarized in Table 4.2; Fig. 4.3 shows a simple hydrogeological scheme for each.

The four springs are the spring of La Villa, which drains the Torcal de Antequera massif (Malaga; Photos 4.1, 4.2 and 4.3); the Fuente Mayor de Simat de Valldigna (Photo 4.4), one of the springs that drains the eastern edge of the massif of Sierra Grossa (Valencia); and the Aliou and Baget springs, both in the French Pyrenees (department of Ardèche). With the exception of Baget, which also has a run-off fraction, discharge from each system is exclusively through its spring. The study hydrographs are shown in Fig. 4.4.

Table 4.2 Main characteristics of the karst systems studied

Aquifer/parameter	Torcal[a]	Simat[ab]	Aliou[c]	Baget[c]
Spring	La Villa	Fuente Mayor	Aliou	Baget
Area km^2	28	20	12	13
Max altitude m	1369	846	1251	1417
Height of spring m	586	60	441	450
Relief	Abrupt	Abrupt	Very abrupt	Very abrupt
Qmax l/s	1785	960	20.000	9.400
Qmin l/s	90	80	10	50
Qmed l/s	411	224	467	530
Average rainfall mm	825	807	1690	1690
Vegetation	Very low	Scarce	Forest	Forest
% Rainfall	56	44	73	76

[a]Soil is very scarce
[b]There is a polje in its basin
[c]There is no karst terrain in this basin, but it is abundant in Baget

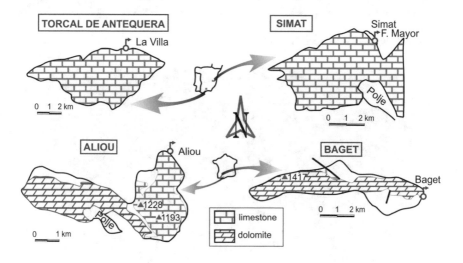

Fig. 4.3 Hydrogeological schemes of the four aquifers

Although a spring's recession is not an ideal study to characterize the response of a system and compare it to other springs, it is representative of the behaviour of an aquifer (in Chap. 1 [18], [11, 14–16]). As we shall see, the recession of each of the springs is very different, which gives a firm incentive to describe their patterns.

Photo 4.3   Weather station next to La Villa spring

Photo 4.4   Pond in the Fuente Mayor de Simat de Valldigna

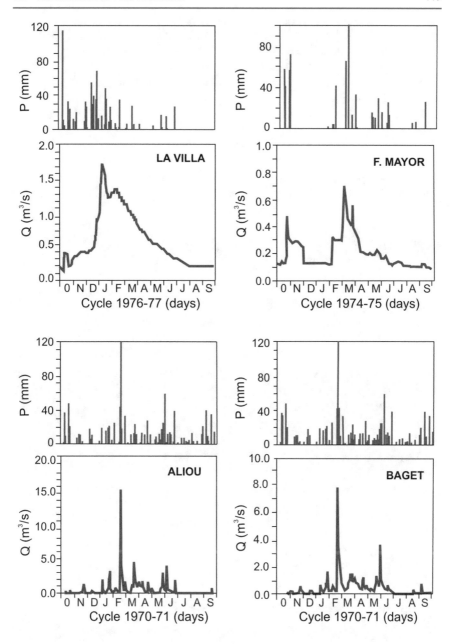

Fig. 4.4 Flow rates and rainfall for one year in the four systems. The same rainfall data are used for Aliou and Baget springs

4.4   Box 1: Application of Classical Methodology

There is only a discontinuous record of the flow from the karst spring at Alomartes, a village in the province of Granada, yet this aquifer was the subject of a thesis and several published articles, one of which [17] serves as the basis for this section. In quantitative terms, Alomartes spring is the main drainage point of the karst system to the south of the Parapanda massif, and its surface area is 24.6 km^2, with a minimum and maximum elevation of 700 and 1604 m, respectively (Photo 4.5). From a geological point of view, it constitutes a klippe of Jurassic calco-dolomitic materials corresponding to the internal Sub-Betic zone, resting on marly and limestone material of the Middle Sub-Betic zone (Fig. 4.5). The exokarst of the Parapanda massif is well-developed.

The carbonate material of the fissured massif is more than 600 m thick. The limits of the aquifer correspond to those of the Sierra de Parapanda, in a geological sense. The southern edge is partly covered by Plio-Quaternary material. Recharge is achieved by the infiltration of rainwater, and occasionally from snowmelt. The average annual rainfall is about 24 hm^3, of which approximately 55% constitutes effective infiltration. The discharge is

Photo 4.5   Panoramic view of the Sierra de Parapanda (*Photo* M.T. Leboeuf)

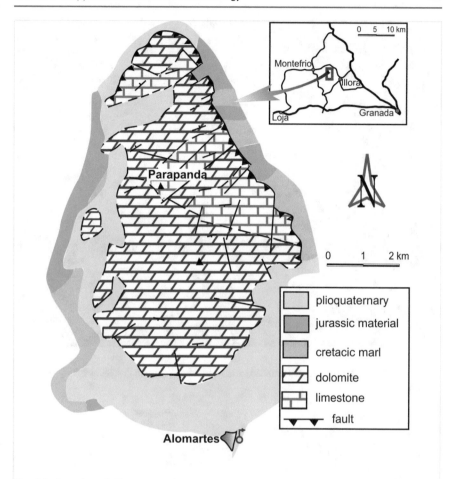

Fig. 4.5 Location of Alomartes spring

through natural surface spillways, and an estimated 70% of the total is drained by its two most important springs, Alomartes (Photo 4.6) and Cerezos.

Figure 4.6 shows the evolution of the Alomartes discharge flows over time, based on monthly observations together with monthly rainfall data from the stations at Íllora, Montefrío and Parapanda. From this graph, it can be seen that the spring is perennial, with flows ranging from just over 500 to 25 l/s. The spring response to rainfall is rapid and pronounced. Some sections of the hydrograph have been decomposed (Fig. 4.7).

Photo 4.6 View of the Sierra de Parapanda and Alomartes spring, in that municipality (*Photo* A. Pulido)

4.4.1 Depletion Graph Analysis

Using 17 flow measurements carried out between March and September, the 1977 hydrograph was drawn with semi-logarithmic coordinates (Figs. 4.8 and 4.9). The final part of the spring's depletion is at a steeper angle than previous sections. This is interpreted as a sign of a lack of continuity in the aquifer at depth, below the level of spring, and of much karstification at that level.

Taking into account the hydrograph components and following Schoël-ler's methodology, the emptying of the aquifer corresponds to the superimposition of two types of underground flow, marked on the hydrograph by the straight lines A and B. The flow at the instant 't' is the sum of the exponentials: $Q_t \ (\mathrm{m^3/s}) = 0.325 \ \mathrm{e0}^{-0.062 \ t} + 0.224 \ \mathrm{e0}^{-0.0045 \ t}$. At first, the depletion progresses rapidly. The type of flow that it represents gradually becomes less important with respect to total flow. By 70 days after the peak flow was reached, this term is negligible and, therefore, the flow at instant 't' is governed almost exclusively by the second term of the equation. Thus, it can be concluded that the spring is fed mostly by groundwater stored in minor discontinuities. The initial volume, above the spring level, is the sum of two volumes:

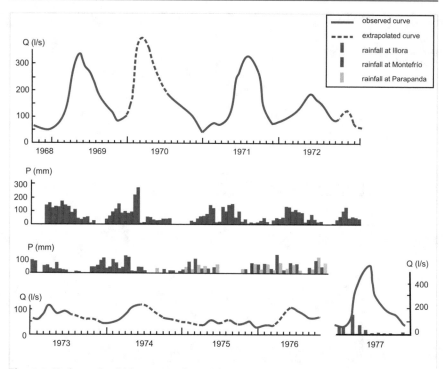

Fig. 4.6 Hydrograph of Alomartes spring, with monthly data

$$V_0\left(m^3\right) = V_0' + V_0'' = (0.325/0.062 + 0.224/0.0045)$$
$$\times\, 86.400 = 48\,hm^3. \tag{4.5}$$

The volume that can be drained by gravity after 120 days of low water ($t = 120$ days) is a function of the flow at that instant, and of the depletion coefficient ($\alpha_2 = 0.0045$): $V_t = (q''/\alpha_2) \times 86.400$ and we arrive at $V_t = 2.5\,hm^3$.

The difference between the initial volume (V_0) and the residual volume after 120 days (V_t) gives the volume of water discharged by the spring between time $t = 0$ and $t = 120$ days, which is 2.3 hm^3. These formulas correspond to various expressions of the general equation, which defines a decrease of flow in time according to a branch of the hyperbola [6] of exponent 'n' equal to 1/2 and 2, respectively:

$$Q_t = \frac{Q_0}{(1 + \alpha_t)^n} \tag{4.6}$$

where Q_t = discharge for a time 't'; Q_0 = discharge for the start of depletion ($t = 0$); and α = coefficient of decrease of flow in time.

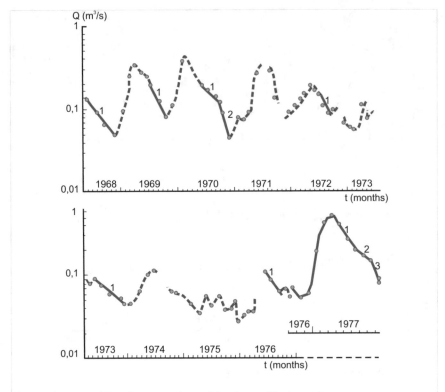

Fig. 4.7 Decomposition of some sections of the observed hydrograph

Fig. 4.8 Hydrograph of
observations of Alomartes
spring, semi-logarithmic
coordinates

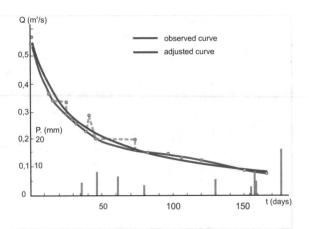

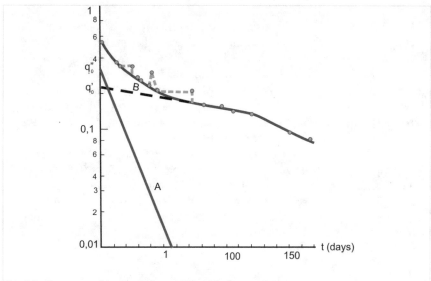

Fig. 4.9 Decomposed hydrograph, semi-logarithmic coordinates

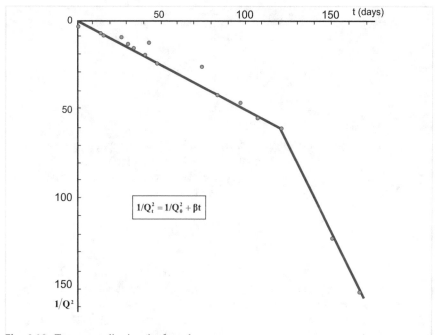

Fig. 4.10 Tests to application the formulas

From the results obtained (Figs. 4.10 and 4.11), it can be deduced that the Forkasiewicz and Paloc formula is not the same for the entire depletion period and therefore has the same disadvantages as the exponential formula. Moreover, Tison's formula is the same for all periods of low water; consequently, the decrease of the flow of Alomartes spring over time responds sufficiently nearly to a branch of the hyperbola defined by the expression: $Q_t = Q_0/(1 + \alpha\ t)^2$. Finally, the principal volumes of depletion and discharge over 120 days are comparable, regardless which formula is used.

Tests by applying the various formulas have shown that the function $Q = f(t)$, in a period of depletion, decreases according to a branch of the hyperbola and that the exponent '$n = 2$' of its general equation represents Alomartes spring's depletion curve quite well. However, a new formula has been tested, one that has been derived from the simplest equation of those defining hyperbola branches, in order to obtain a better fit. This is how the equation could be established:

$$\frac{1}{Q_t} = \frac{1}{Q_0} + \beta_t \tag{4.7}$$

where 'β' has the meaning of a coefficient of depletion flow in time. This equation, in Drogue's generalization, corresponds to that of exponent 'n' equal to one. The integration of (4.1) between two given instants allows the calculation of the volume of water drained in that interval:

$$V = \frac{Q_0}{\alpha}[\ln(1 + \alpha t) - \ln(1 + \alpha\ t_0)] \tag{4.8}$$

If $t_0 = 0$: V_e (m³) = $Q_0/\alpha\ \ln(1 + \alpha t) \times 86.400$, where '$\alpha$' is Drogue's depletion coefficient of $Q = f(t)$.

From the application of this new formula for Alomartes spring's depletion, by linear correlation by least squares, gives the line shown in Fig. 4.12, with the equation: $1/Q_t = 1.884 + 0.057\ t$ (correlation coefficient $r = 0.9851$).

According to Drogue's generalization, the equation that governs depletion is:

$$Q_t = \frac{0,54}{1+0,03t} \tag{4.9}$$

The volume of water drained by the spring, from the start of depletion to the instant $t = 120$ days, is 2.4 hm³. The proposed formula has the advantage of remaining valid for the entire depletion period; moreover, it is easier to use than the exponential formula or any of the above methodologies. The validity of the expression (4.9) as a characteristic formula of the depletion period of Alomartes spring must be sought in the variation of the coefficient β from one year to the next.

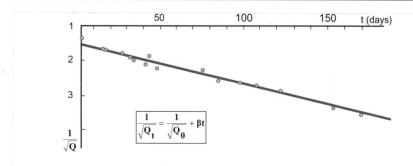

Fig. 4.11 Tests to application the formulas

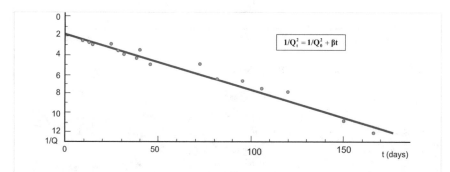

Fig. 4.12 Representation of the proposed formula

As a synthesis, and to provide the hydrodynamic influences on the aquifer as deduced by analysis of the entire depletion curve, from the proposed expression we have calculated the theoretical discharge as a function of time. The representation is by arithmetic coordinates of the adjusted depletion curve (Fig. 4.12), showing the close proximity of the adjusted curve and the observed depletion curve (mean deviation of 0.015 m^3/s). The decomposition of the adjusted hydrograph (Fig. 4.13) indicates the presence of three components of underground flow, shown on the hydrograph by three lines at an angle to each other, so that after reaching peak flow (Q_0), the flow at instant 't' is the sum of three exponentials [18]:

$$Q_t(m^3/s) = 0.09e^{-0.2t} + 0.27e^{-0.04t} + 0.24e^{-0.0062t} \qquad (4.10)$$

The volume of water stored above the level of spring, which is drainable by gravity, at instant t_0 is the sum of the three volumes corresponding to the three categories of groundwater flow:

$$V_0(m^3) = V'_v + V''_0 + V'''_0 = (q'_0/a_1 + q''_0/a_2 + q'''_0/a_3) \times 86.400.$$

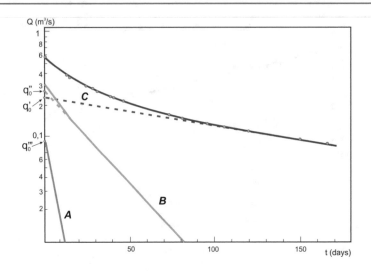

Fig. 4.13 Decomposition of the adjusted hydrograph

For the depletion curve being studied, the respective volumes are $V_0' = 42,492$ m^3; $V_0'' = 580,298$ m^3; and $V_0''' = 3,366,233$ m^3, which represents an initial volume (V_0) of approximately 4 hm^3.

Although in karst springs the flow curve as a function of residual volume (Fig. 4.14) cannot be considered to be a characteristic element, it makes it possible to visualize the relationship between the decreases in flow and the residual volume. Thus, a decrease is observed according to a parabola branch' in such a way that, at the beginning of the depletion, the decrease in the residual volume is rapid.

For the depletion analysed, the initial volume is made up of approximately 84% of the water stored in the type of fissures, producing the component 'α_3', while the components 'α_2' and 'α_3' represent 14 and 1%, respectively, of the total water stored (Table 4.3). With regard to the speed of emptying the aquifer, the curve of the released capacity $V_d = f(t)$ in Fig. 4.15 indicates rapid emptying in the first 20 or 30 days. Indeed, as seen below, the correct terminology is *quick flow* and *base flow* $V_0 = V_0' + V_0'' + V_0'''$.

The results make it possible to determine the presence of three decreasing speeds: 'α_1' corresponds an emptying 33 times faster than 'α_3', and 'α_2' corresponds to a less rapid emptying (6.5 times faster than for 'α_3') and, finally, very slow emptying. The analysis of the volumes shows the importance of smaller discontinuities to water storage: the role played by microfissures is fundamental.

It is interesting to ascertain the mathematical expression to define the entire hydrograph of a spring [7]. Testing a global mathematical expression yields some interesting conclusions, in this case. From the mathematical

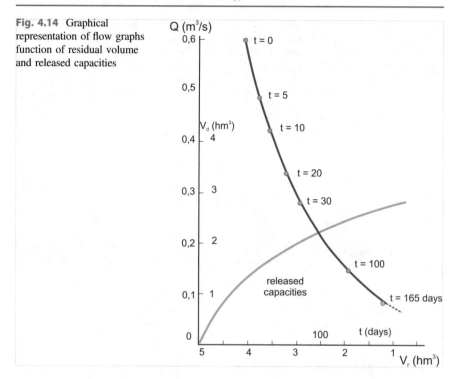

Fig. 4.14 Graphical representation of flow graphs function of residual volume and released capacities

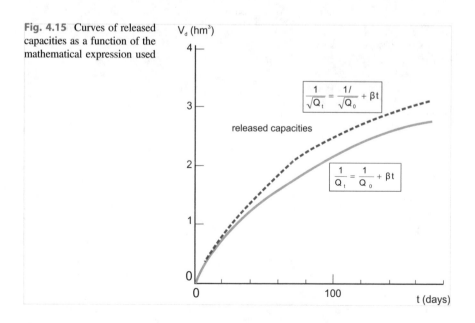

Fig. 4.15 Curves of released capacities as a function of the mathematical expression used

Table 4.3 Q_0: Observed peak flow; Q_0': Theoretical peak flow; α_1, α_2 and α_3 = base flow coefficients of straight lines A, B and C, respectively. V_0', V_0'' and V_0''' = respective volumes of straight lines A, B and C. V_0 = initial volume; V_d = volume drained in 120 days

$Q_0 = 0.54$ m^3/s	$V_0' = 0.04$ hm^3		$V_0'/V_0 = 1\%$
$Q_0' = 0.56$ m^3/s	$V_0'' = 0.6$ hm^3		
$\alpha_1 = 0.21$	$V_0''' = 3.4$ hm^3	$V_d = 2.4$ hm^3	
$\alpha_2 = 0.04$	$V_0 = 4$ hm^3	$V_0''/V_0 = 84\%$	
$\alpha_3 = 0.0062$		$V_0'''/V_0 = 14\%$	

expression proposed by Galabov and by application of the graphical method, we have arrived at Eq. (4.11), considering a permanent minimum discharge (q) of 55 l/s:

$$Q'(\text{m}^3/\text{s}) = Q - q = 0.08x\frac{t^{3.36}}{e^{0.024t}} \tag{4.11}$$

or:

$$Q(l/s) = 55 + 0.0009x\frac{t^{3.36}}{e^{0.024t}} \tag{4.12}$$

The graphical representation of Eq. (4.12) is shown in Fig. 4.16. It can be observed that this expression faithfully translates the hydrograph's increasing branch, yet not its decreasing one. It follows that it is not possible to adjust it by means of Galabov's expression, in this case, given that it cannot accept asymmetries in the hydrograph as marked as those studied here. This is very frequent in karst hydrographs, apart from in the early stages of base flow.

Fig. 4.16 Hydrograph of Alomartes spring 1976–77, according to Galabov's expression

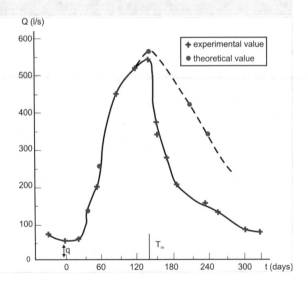

4.5 Mangin's Equation

This method is based on distinguishing two subsystems in the karst system: the *infiltration subsystem*, corresponding to the unsaturated zone in which variations in flow (q) is responsible for the floods and decreases observed in a spring; and the *saturated subsystem*, corresponding to the saturated zone, and whose outflow (through a spring) is Q. According to Mangin (in Chap. 1 [18],) any increase in flow (flooding) observed in a spring is indicative of $q > Q$. By contrast, in the decreasing stage $q < Q$ and in peak discharge it is $q = Q$. As soon as infiltration ceases ($q = 0$, uninfluenced system), depletion, or base flow, is considered to have begun.

The recession curve can be broken into two parts. One is non-exponential, where the flow decreases very rapidly and represents the decrease itself; and the other is approximately exponential, where the flow decreases slowly and represents depletion. The shape of the curve is a direct consequence of the state of the saturated zone (depletion) and the way in which the infiltration happens (decrease).

Mangin (in Chap. 1 [18]) considers that during the recession of a karst spring the flow at a time t responds to the expression:

$$Q_t = \phi_t + \psi_t \tag{4.13}$$

where Ψ_t is an infiltration function that transfers the effects of recharge through the unsaturated zone to the spring, modified to some extent by its passage through the saturated zone; and ϕ_t would be Maillet's formula [1]:

$$\phi_t = q_t^b = q_0^b e^{-\infty t} \tag{4.14}$$

where q_t^b is the base flow at time t; q_0^b is the base flow extrapolated at the beginning of the recession, as shown in Fig. 4.17; and α is the depletion coefficient (of the base flow).

Fig. 4.17 Hydrographs of the four recessions chosen for analysis

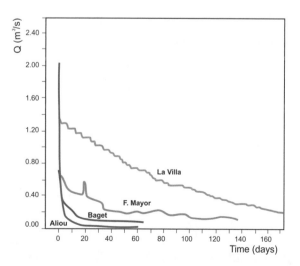

Ψ_t is an empirical function that is expressed as:

$$\psi_t = q_t^* q_0^* \frac{1 - \eta t}{1 + \varepsilon t} \qquad (4.15)$$

where q_t^* is the rapid flow at time t; q_0^* is the difference between the flow Q_0 at time $t = 0$ and the base flow component, q_0^b; and is η $1/t_i$. The function is defined between $t = 0$ and ti ($t_i = 1/\eta$), which is the duration of the fast flow. The coefficient ε characterizes the importance of the concavity of the fast flow curve; the values of this parameter allow to estimate the moderating capacity of the unsaturated zone on the infiltration. Mangin also defines the function $Y_t = \Psi_{t/} q_0^*$, which, according to (4.15), remains:

$$Y_t = \frac{1 - \eta t}{1 + \varepsilon t} \qquad (4.16)$$

Y_t varies between 0 (for $t = t_i$) and 1 (for $t = 0$) independent of the input, which makes it useful for comparing the quick flow of springs. If Q_t, q_{tb} and q_t^* are the total flow, the base flow and the rapid flow, respectively, drained by the spring in time t, we arrive at the expression:

$$Q_t = q_t^b + q_t^* q_0^b e^{-\alpha y} + \frac{1\eta}{1 + \varepsilon t} \qquad (4.17)$$

where q_0* is estimated from the difference between Q_0 and q_0^b (Fig. 4.18), and Q_0 is the total flow at the start of the recession. The parameter ε coincides with the slope of the line that best matches the slope of the Z_t representation with respect to t:

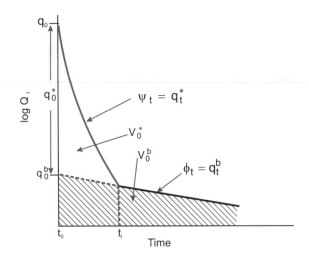

Fig. 4.18 Idealized diagram showing quick flow and base flow in the recession, according to Mangin's equation. The text explains the variables

$$Z_t = \frac{q_0^*}{q_t^b}(1 - \eta t) \tag{4.18}$$

The initial volume of stored water V_0^b in $t = 0$ capable of being drained by base flow can be calculated from the expression

$$V_0^b = \int_0^\infty q_0^b e^{-\alpha t} dt \frac{q_0^b}{\alpha} \tag{4.19}$$

The volume drained during the base flow from the start of the recession for time $t = 0$ is:

$$V_t^b = \int_0^t q_0^b e^{-\alpha t} dt = \frac{q_0^b}{\alpha}(1 - e^{-\alpha t}) \tag{4.20}$$

The volume of water initially stored (V_0^*) that can be drained during the period of rapid flow can be calculated by integrating the expression $v_0^* = \int_0^{ti} q_0^* \frac{1 - \eta t}{1 + \varepsilon t} dt \frac{q_0^*}{\varepsilon}$, obtaining:

$$\left[\mathrm{Ln}(1 + \varepsilon t)\left(1 + \frac{\eta}{\varepsilon}\right) - 1 \right] \tag{4.21}$$

The volume drained by the spring at time t ($0 < t < t_i$) of quick flow is:

$$V_t^* = \int_0^t q_0^* \frac{1 - \eta t}{1 + \varepsilon t} dt = \frac{q_0^*}{\varepsilon} \tag{4.22}$$

The initial total volume (V_0) that the spring will provide is given by the sum of the initial dynamic volume of the base flow, and the initial volume that it will discharge as quick flow: $V_0 = V_{0b} + V_0^*$. By dynamic volume, we mean the volume that can be drained by gravity through a spring in time $t = 0$, the beginning of the recession.

According to this author, the method can be used to classify a karst system. The expression $\Psi(t)$ is characterized by two parameters. One is η = 'infiltration velocity coefficient', which is the duration of infiltration. The bigger η value, the shorter is the time of influence of the infiltration; that is, the faster the infiltration. Its dimension is T^{-1}. The other is ε = 'heterogeneity coefficient', which is the importance of the concavity of the curve. When ε is large, it indicates a smoothing of the decreased flow after a very fast decrease of the flow in a spring, created by an also rapid decrease in its infiltration flow. Its dimension is T^{-1}. $\varepsilon = \frac{1}{t} - 2\eta$ where *t* has been chosen as the time that must elapse for $q = q_0/2$. The determination of ε and η allows to solve the function **y** $y = \frac{\psi(t)}{q_0} = \frac{1-\eta t}{1+\varepsilon t}$.

This function is independent of the amplitude of the flood, and its value oscillates between 0 and 1. For the same system, there is a family of functions **y**, relatively close together, which serve to differentiate the various behaviours, for example on the basis of the type of contribution (snow, rain, etc.), characterizing the system under investigation and acting as a criterion for comparison with other systems. The function **and** represents, in a way, how the input function (rainfall) has been modulated as it passes through the system, and this may depend on each situation's specific circumstances. Based on the hydrodynamic characteristics of a system (function and depletion coefficient, etc.), Mangin (in Chap. 1 [18]) proposed the establishment of parameters that, in addition to characterizing the system, allow its comparison to other systems. These parameters are the power regulator (*k*) and the index i.

Regulatory power (k) is the capacity of the system to progressively restore part of the volume of water that passes through it (transit volume). It is defined as:

$$k = \frac{V_d}{V_t} \tag{4.23}$$

where V_d represents the dynamic volume—the longer the observation period, the greater its representativeness; and V_t is the volume that passes through the system during an average hydrological year. In this way, a system capable of restoring the totality of its transit volume in the form of base flow, as described by Maillet's law, will have maximum regulatory power ($k = 1$). By contrast, if it is incapable of restoring its reserves outside of flood periods—that is to say, if it has no base flow—its regulatory power will be null. In the first instance, it can be seen that karst aquifers are in the realms of $k < 0.5$, and that regulatory power is weak in systems for which $k < 0.1$, with predominantly rapid types of infiltration (in aquifer systems) and surface run-off (in surface systems).

Index i characterizes a system on the basis of the shape of its mean decreasing curve. Mangin proposed using the value *i* in the function **y** for $t = 2$ days. The choice of this value of *t* lies in the fact that it is sensitive to the various graphs of the systems, since at times of greater flow (*t*) the values of *i* are grouped excessively tightly. Based on the study of various karst systems, Mangin estimated that the systems for which $i > 0.5$ are complex and large. If $0.25 < i < 0.5$, the system experiences a delay in receiving any contribution, while for $i < 0.25$ the system has a highly developed drainage network.

Based on the knowledge of these two parameters (*k*, *i*), a classification (Fig. 4.19) of the five domains can be constructed. Domain I corresponds to highly karstified systems, especially the saturated subsystem. Domain II is similar, but with an important saturated zone (greater power of regulation). In Domain III, the karstification is greater in the subsystem infiltration, with gaps in the water feed (due to the presence of non-karst terrain, snow, etc.). Domain IV describes complex systems, while Domain V has very few or no karstified systems.

Fig. 4.19 Systems drained by karst springs: A: Aliou, AR: Argin; B: Baget, F: Fontestorbes; O: Orue; V: Vaucluse; FL: Fontaine l'Evêque [13, 14]

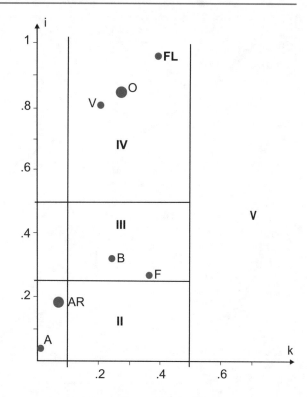

These functions are also applicable to a spring's hydrographs and have been used with the springs at La Villa, Fuente Mayor, Aliou and Baget.

4.5.1 Box 2: Application of Mangin's Equation

Figure 4.20 is a graphical representation of the Ψ_t function of quick flow, and the Φ_t function of base flow. It follows that the spring of La Villa shows an absence of quick flow during its period of recession. In this case, the only expression used is that of Maillet. From the values of the parameters obtained (Table 4.4), we can draw the following conclusions about its quick flow. This adaptation of the Ψ_t function is acceptable for the spring at Aliou but less so for Fuente Mayor and Baget, where it is necessary to extrapolate the curve to allow for the presence of minor flow deviations. However, we consider that the values of the parameters η and ε characterize these curves adequately.

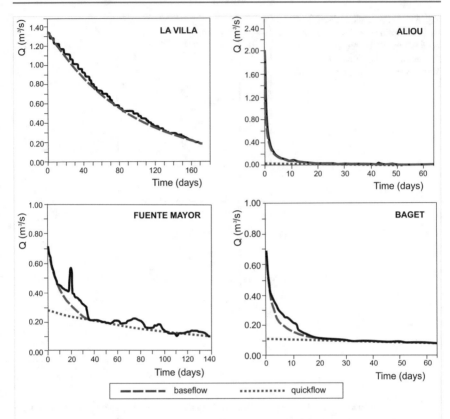

Fig. 4.20 Total discharge observed and base and quick flows estimated from Mangin's expression of recession in the four karst springs

Table 4.4 Synthesis of the main parameters characterizing the recession curves

Parameter/spring	La Villa	Fuente Major	Aliou	Baget
Q_0 (m³/s)	1.350	0.71	2.02	0.7
q_0^b (m³/s)	1.350	0.275	0.045	0.113
α	0.013	0.007	0.025	0.006
q_{0*} (m³/s)	–	0.345	1.975	0.587
t_i (days)	–	36	32	34
η	–	0.028	0.032	0.029
ε	–	0.084	2.63	0.7
$V_0^b/V_{0\%}$	100	90	40	91
$V_0^*/V_{0\%}$	–	10	60	9

Figure 4.21 shows the values of the Y_t function. The differences are mainly due to the springs' ε value (characterizing quick flow). The unsaturated zone of the system drained by the Fuente Mayor ($\varepsilon = 0.0844$) plays a greater role in the flow response than at Aliou ($\varepsilon = 2.63$). Baget ($\varepsilon = 0.7$) lies between the two, although its characteristics are closer to Aliou's. This difference is seen clearly in Fig. 4.22, which shows the percentage ratio of the volume drained at time t and the 'volume stored' at time $t = 0$, for quick flow ($V_t^*/V_0^* \%$).

Most of the water supplied by Fuente Mayor and Baget is drained as base flow (90% and 91%, respectively); the rest is quick flow (10 and 9%), and at Aliou, it is 40 and 60%. Figure 4.22 shows the drained volume compared to the initial volume—as a percentage—for quick flow ($V_t^*/V_{0\%}$), base flow ($V_t^b/V_{0\%}$) and the sum of both ($V_t/V_{0\%}$) for the springs being investigated. Table 4.3 shows the parameter values obtained for the functions of the recession periods studied.

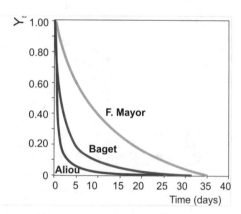

Fig. 4.21 Change of function Y_t function in time for quick flow for the three karst springs (La Villa has only base flow ($Y_t = 0$))

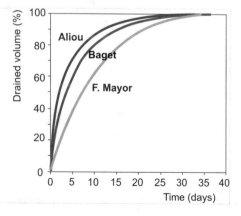

Fig. 4.22 Percentage of volume drained to stored volume, for quick flow only

From this, we can conclude that the saturated zone of the aquifer of El Torcal de Antequera supplies almost the entire flow of the spring of La Villa; moreover, it does so at a higher speed than at Simat or Baget. In other words, it suggests that El Torcal de Antequera aquifer is highly transmissive. While the saturated areas of Simat and Baget are well-developed in terms of their volume (90 and 91% of the respective totals), their recharge speed is low. By contrast, the saturated zone at Aliou is poorly developed, yet it is highly transmissive due to its network of karst conduits, and its discharge after rainfall is very rapid.

These findings are corroborated by calculation of the volume drained during base flow as a proportion of the total susceptible to the spring's drainage (Fig. 4.23). For example, by 25 days after the onset of the recession period, La Villa discharges 25% of the draining volume, Fuente Mayor 15%,

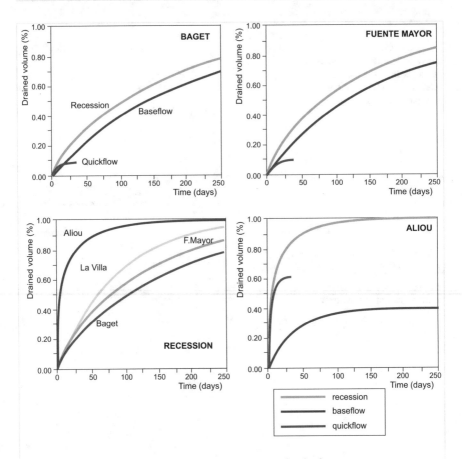

Fig. 4.23 Percentage of volume drained to stored volume, for the four systems

Aliou 20% and Baget 13%. The absence of quick flow at La Villa indicates that the water does not access the saturated zone quickly. This is probably due to poor development of the karst network in the unsaturated area. In the other three karstic systems, there is a well-defined quick flow curve, suggesting that water reaches the saturated zone rapidly through discontinuities in the unsaturated band. It should be noted that the flow from the Simat aquifer is more homogeneous than at Baget. At Aliou, the infiltration of the precipitation is controlled essentially in the unsaturated zone, flowing through highly conductive zones that transmit water to the spring. During the same 25-day period as discussed above, for the saturated zone, the percentage of water drained as quick flow is 10% for Fuente Mayor, 9% for Baget and 60% for Aliou.

4.6 Coutagne's Equation

4.6.1 Theoretical Development

The recession of a spring's hydrograph can be regarded as a continuous function, as in the drainage of a reservoir. One of the best functions adapted to karst spring hydrographs is Coutagne's equation [11]:

$$Q = CV^n \qquad (4.24)$$

where Q is the flow; C is a constant; V is the stored volume available for drainage by the spring; and n is an exponent whose value varies between 0 and 2. The flow in any time Q_t is:

$$Q_t = dV/dt \qquad (4.25)$$

where

$$dV/dt + CV^n = 0 \qquad (4.26)$$

The solutions to this differential equation are:

a. for $n = 1$:

$$Q_t = Q_0 e^{-\alpha t} \qquad (4.27)$$

where Q_0 is the flow at the beginning of the recession; α it is the recession coefficient; and t is time.

b. for $n1 \neq$:

$$Q_t = Q_0[1 + (n-1)\alpha_{0t}]^{n/(1-n)} \tag{4.28}$$

where α_0 is the recession coefficient for Q_0, i.e. for time $t = 0$.

Coutagne's equation represents a general case of recession in hydrographs, and its physical significance depends on the value of n. When $n = 0$, the aquifer is emptied at a constant flow from the outlet; $n = 1$ corresponds to a situation similar to that in Maillet's equation [1; Eq. (4.2)] and reflects discharge at constant velocity; when $n = 2$, the solution represents an aquifer that discharges in a laminar regime, with the drainage proportional to the amount of water stored in the system. In this case, the flow is proportional to the square of the hydraulic head. The range of values of n is very wide; hence, the choice of graphs to which this expression can be adapted makes it very useful in the analysis of the hydrographic recession of a karst spring.

At time $t + \Delta_t$ in the recession curve, the flow is:

$$Qt + _{\Delta t} = Q_0\left[1 + (n1-)\alpha_0(t + \Delta t]^{n/(1n-)} \tag{4.29}$$

If we divide (4.28) by (4.29), we get:

$$\frac{Q_{t+\Delta t}}{Q_t} = \left[1 + \frac{(n-1)\alpha_0(\Delta t)}{1 + (n-1)\alpha_0 t}\right]^{n/(1-n)} \tag{4.30}$$

Expressed otherwise:

$$Q_t + _{\Delta t} = Q_t[1 + (n1-)\alpha_t\Delta]^{n/(1n-)} \tag{4.31}$$

in which α_t is

$$\alpha_t = \frac{\alpha_0}{1 + (n-1)\alpha_0 t} \quad \text{o} \quad \alpha_t = \alpha_0\left(\frac{Q_0}{Q_t}\right)^{(1-n)/n} \tag{4.32}$$

It follows that the recession coefficient α_t is variable in time depending on α_0, n, and flow in time t.

The volume stored at time t (V_t) is :

$$V_t = \int_t^\infty Q_0[1 + (n-1)\alpha_0 t]\mathrm{d}t \tag{4.33}$$

whose integration [16] is:

$$V_t = Q_t/\alpha_t \tag{4.34}$$

Dividing by α_t and α_0 gives a new H_t function that does not depend on the input peak and is similar in nature to the Y_t function described by Mangin yet refers to the entire recession. This function varies between 1 (for $t = 0$) and 0 (for $t = \infty$) and has the form:

$$H_t = \frac{1}{1 + (n-1)\alpha_0 t} \tag{4.35}$$

Coutagne's equation is not applicable to the study of karst springs with a frequency $n < 1$, as these aquifers have a pronounced recession. In these cases, the function decreases rapidly and reaches negative values. This is why it is necessary to introduce a new Q_c parameter for discharges with very little variation over time. This flow may come from the discharge of aquitards in contact with the karst aquifer or from the discharge of areas of the aquifer that are not highly transmissive. The new expression for Q_t is:

$$Q_t = (Q_0 Q_c)[1 + (n-)\alpha_{0t}]^{n/(1n-)} + Q_c \tag{4.36}$$

and the stored volume to be drained by the aquifer at time t (V_t) is:

$$V_t = \frac{Q_t - Q_c}{\alpha_t} \tag{4.37}$$

The advantage of this variant is that it allows Coutagne's equation to be adapted to the entire recession of any karst spring.

4.6.2 Box 3: Application of Coutagne's Equation

Coutagne's equation was applied to the four previous hydrographs, and the results are shown in Fig. 4.24; in La Villa, the result is the same, since base flow is $n = 1$. The parameters of the function are shown in Table 4.5. The values of n suggest that the relationship between the flow of the springs and the hydraulic load is linear in La Villa, while in the other three it is proportional to h^n.

$$H_t = \frac{1}{1 + (n-1)\alpha_0 t} \tag{4.38}$$

Fig. 4.24 Flow rates
observed and simulated using
the Coutagne equation for
karst aquifers

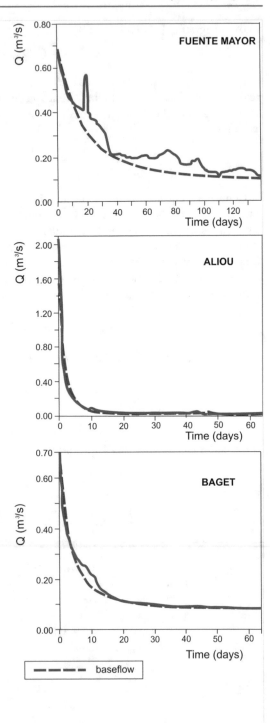

Table 4.5 Parameters adjusted to the Coutagne equation for the four recessions studied (Q_0 and Q_c in m³/s, α_0 in days⁻¹)	Spring	Q_0	α_0	n	Q_c
	La Villa	1.35	0.013	1	–
	Fuente Major	0.75	0.040	1.5	0.09
	Aliou	2.02	0.330	2	–
	Baget	0.70	0.160	2	0.08

The H_t function, which describes the aquifer's discharge, has been calculated from the hydrographs of the springs at Fuente Mayor, Aliou and
Baget (Fig. 4.25). At Aliou and Baget, there is a sharp jump at the start of the
recession, followed by a period rapid change towards stability; in Fuente
Mayor, the initial fall is less pronounced and stabilization takes longer than in
the other two springs. This fact is also reflected in Fig. 4.26, which shows the
percentage of the volume drained by the springs of to the amount of initial
water stored in the system. Comparing the function H_t with Y_t in Fig. 4.27
shows similar behaviour, although the latter includes total recession while the
other involves only base flow.

If we consider the recession curve from the discharge of the saturated zone
of the aquifer, we can deduce that the aquifers of Aliou and Baget have a
highly karstified saturated zone with large drainage conduits that drain
quickly but have a reduced storage capacity. In Simat, the initial discharge is
less rapid, due to the possible absence of large conduits, with the predominant
diffuse flow being fairly homogeneous across the aquifer mass. The Torcal
aquifer is an extreme case, with a flow more akin to that of an intergranular
porosity medium than a karst aquifer [15], with a constant recession coefficient independent of the hydraulic head.

Figure 4.27 shows the variation in the coefficient α_t as a function of the
spring flow, calculated from Eq. (4.18). For relatively high flows, Aliou has
the highest coefficient, followed by Baget, while Fuente Mayor has the

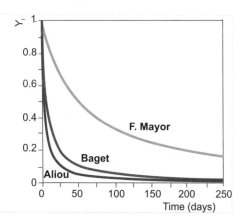

Fig. 4.25 H_t change over time for karst springs during recession

Fig. 4.26 Percentage of the volume drained since the start of each spring's recession

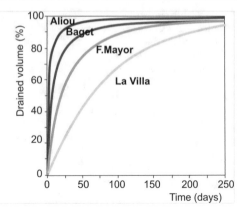

Fig. 4.27 Variation of the coefficient with respect to the flow rate for the springs studied

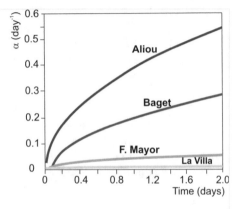

lowest. In the first two, there are appreciable changes in flows of less than 0.3 m³/s, while in Fuente Mayor the α variations with the flow are relatively minor for low flows and almost constant for high flows.

4.7 Further Reading

Boussinesq, J. 1904. Recherches théoriques sur l'écoulement des nappes d'eau infiltrée dans le sol et sur le débit des sources. *Journal de Mathématiques Pures et Appliquées*, 10: 5–78.

Grund, A. 1903. Die Karsthydrographie. Studien aus Westbosnien. *Geogr. Abhandlungen von A. Penck*, VII (3): 1–201.

European Commission. 1995. *Karst Groundwater Protection*. COST Action-65. Report EUR 16547 EN, 446 p.

Fiorillo, F. 2014. The recession of spring hydrographs, focused on karst aquifers. *Water Resources Management*, 28: 1781–1805.

Maillet, E. 1905. *Essais d'hydraulique souterraine et fluviale.* Paris: Hermann, 218 p.
Mangin, A. 1984. Pour une meilleure connaissance des systèmes hydrologiques à partir des analyses corrélatoire et spectrale. *Journal of Hydrology,* 67: 25–43.

4.8 Short Questions

1. Design a gauging station for a karst spring.
2. Design a gauging station for a karst polyemergency, in both a vertical and a horizontal plane.
3. To characterize a karst spring adequately, why is it useful to have data on the daily discharge?
4. What does Maillet's formula establish for the depletion of karst springs?
5. What is a spring recession, and how it can be studied?
6. Briefly discuss Schoëller's method for studying the depletion of a karst spring.
7. Discuss Drogue's method for studying depletion in a karst spring.
8. Do you think that Galabov's method has good application to the study of karst springs? Give reasons for your answer.
9. Briefly explain the difference between *quick flow* and *base flow.*
10. Discuss the most relevant differences between the springs at La Villa, Baget and Aliou.
11. Explain Mangin's equation and give his mathematical expression.
12. In Mangin's terminology, what is transit flow, and how is it calculated?
13. Briefly explain Coutagne's method and its application to the characterization of karst springs.
14. To what extent do analyses of depletion graphs explain the organization of flow within a karst massif?
15. Do you think that topography s reflected by the behaviour of karst springs in both areas of high mountains with complex geology and isolated karst aquifers?

4.9 Personal Work

1. Methods of decomposition of hydrographs of karst springs.
2. Analysis of depletion in karst hydrographs.
3. Apply Galabov's method to the study of karst springs.
4. Identify *quick flow* and *base flow* in karst springs and their hydrodynamic implications.

References

1. Maillet, E. (1905). *Essais d'hydraulique souterraine et fluviale* (p. 218). Paris: Hermann.
2. Boussinesq, J. (1904). Recherches théoriques sur l'écoulement des nappes d'eau infiltrée dans le sol et sur le débit des sources. *Journal de Mathématiques Pures et Appliquées, 10,* 5–78.

3. Wenner P. W., & Sunquist, N. J. (1951). On the ground water recession curve for large water-shade. IAHS, pp. 202–212. Bruselas.

4. Schoeller, H. (1965). Hydrodynamique dans le karst (écoulement et emmagasinement). *Hydrogéologie des roches fissurées.* IAHS-UNESCO Coll. Hydrol. des Roches Fisurées, Dubrovnik. 1, 320.

5. Forkasiewicz, J., & Paloc, H. (1965). Le régime de tarissement de la Foux de la Vis. Etude préliminaire. *AIHS Coll. Hydrol. des Roches Fissurées*, Dubrovnik, 1, 213–228.

6. Drogue, C. (1972). Analyse statistique des hydrogrammes de décrues des sources karstiques. *Journal of Hydrology, 15*, 49–68.

7. Galabov, M. (1972). Sur l'expression mathématique des hydrogrammes des sources et le pronostic du débit. *Bulletin BRGM. Paris, 2*, 51–57.

8. Mangin, A. (1981). Utilisation des analyses corrélatoire et spectrale dans l'approche des systèmes hydrologiques. *Comptes Rendus Académie des Sciences de Paris, 293*, 401–404.

9. Mangin, A. (1981). Apports des analyses corrélatoires et spectrales croisées dans la connaissence des systèmes hydrologiques. *Comptes Rendus Académie des Sciences de Paris, 293*, 1011–1014.

10. Pulido-Bosch, A. (2007). *Nociones de Hidrogeología para ambientólogos* (p. 490). Almería: University of Almeria.

11. Coutagne, A. (1968). Les variations de débit en période non influencée par les précipitations. *La Houille Blanche*, pp. 416–436.

12. Pulido-Bosch, A., & Padilla, A. (1990). Evaluation des ressources hydriques de l'aquifère karstique du "Torcal de Antequera" (Málaga, Espagne). *Hidrogeología, 5*: 11–22.

13. Mangin, A. (1984). Pour une meilleure connaissance des systèmes hydrologiques à partir des analyses corrélatoire et spectrale. *Journal of Hydrology, 67*, 25–43.

14. Antiguedad, I. (1987). *Estudio hidrogeológico de la cuenca del Nervión-Ibaizabal. Contribución a la investigación de los sistemas acuíferos kársticos* (p. 338). Thesis Doct., University of Basque Country.

15. Mangin, A., & Pulido–Bosch, A. (1983). Aplicación de los análisis de correlación y espectral en el estudio de los acuíferos kársticos. *Tecniterrae, 51*, 53–65.

16. Padilla, A. (1990). *Los modelos matemáticos aplicados al análisis de los acuíferos kársticos* (p. 267). Tesis Doct., University of Granada.

17. Casares, J., Fernández–Rubio, R., & Pulido-Bosch, A. (1979). El manantial de Alomartes en régimen de agotamiento (provincia de Granada). Análisis de hidrogramas de surgencias kársticas. Hidrogeol. y Rec. Hidrául. V: 19–36. Pamplona.

18. Tison, B. (1960). Courbe de tarissement. Coefficient d'écoulement et perméabilité du bassin. Mémoires AIHS Helsinki, 229–243.

Time Series Analyses

5

5.1 Glossary

A *correlogram* shows the *autocorrelation coefficients* r_k ($k = 0, 1, 2, \ldots$), after discretization of the time series ($x_1\ x_2, \ldots, x_n$). An optimum is obtained in the results for $m \leq n/3$; for higher values of m (called *truncation value or truncation*), numerical instabilities may appear.

The *energy spectral density* of a stationary random function is obtained from the Fourier transform of the autocorrelation function (the Wiener–Khinchin theorem). It represents the distribution of variances at the various frequencies.

A *cross-correlogram* is calculated in the same way as a simple correlogram and with the same truncation value m. The difference is that the correlation is made between two discretized time series, considering the first ($x_1, x_2, \ldots, x_m$) as the cause of the second ($y_1, y_2, \ldots, y_m$).

The *asymmetry of a cross-correlogram* for positive and negative k-values forces the spectrum to be expressed by a complex number: *spectrum* $(f) = k_{x,y}$ (f) i $q_{x,y}$ (f), where f is the frequency ($f = j/2\ m$) and i is the imaginary number. $k_{x,y}$ (f) is called co-spectrum.

The *complex expression of the spectrum* is $(f) =$ $\big|$ S_x, and (f) $\big|$ cross-amplitude function; exp $[ix, y(f)]$, the phase function. $C_{X,Y}$ is the coherence function. $g_{x,y}$ (f) is the gain function.

A *trop plein* is a typically seasonal karst spring, generally located higher than the mainspring, which comes into being only in periods of intense rainfall.

5.2 Introduction

Hydrographs of springs have traditionally been the starting point for the study of karst aquifers. Mangin showed that a hydrograph can be considered as a time series (Fig. 5.1) that is able to conserve information on the parameter that generates it

A. Pulido-Bosch, *Principles of Karst Hydrogeology*, Springer Textbooks in Earth Sciences, Geography and Environment, https://doi.org/10.1007/978-3-030-55370-8_5

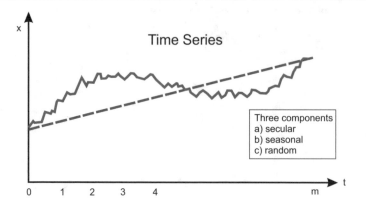

Fig. 5.1 Time series identifying three possible components

(rainfall) at the same time as providing information on the aquifer through which the water circulates. From there, a methodology can be chosen to permit working with a series for as long as possible, not just in recessions but over the entire hydrograph of a rainfall event. On this basis—and much faster—techniques can be used that are traditionally applied to the analysis of surface water.

5.3 Correlation and Spectral Analysis

5.3.1 Simple Analysis

Methods based on the analysis of time series [1, 2] provide a good approximation of the structure and operation of a hydrological system. These methods involve correlation and spectral analyses for inferential purposes, especially in surface hydrology—previsions, filling the gaps in the time series—although they also describe the series and identify their structure and components and thus deduce the system's mechanism of operation.

After discretization of the time series $(x_1, x_2, \ldots, x_n)$, a correlogram corresponds to the autocorrelation coefficients r_k $(k = 0, 1, 2, \ldots)$. Optimum results are obtained for $m \leq n/3$; for higher values of m (called *truncation value or truncation*), numerical instabilities may appear. One of the possible expressions to calculate r_k is:

$$r_k = \frac{C_k}{C_o}; \quad C_k = \frac{1}{n}\sum_1^{n-k}(x_{i+k} - \bar{x}) \tag{5.1}$$

The energy spectral density of a stationary random function is obtained from the Fourier transform of the autocorrelation function (Wiener–Khinchin theorem; [3]). This spectral density represents the distribution of variances for the various frequencies. Its expression is (Fig. 5.2):

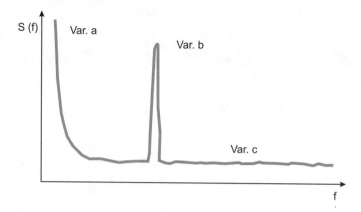

Fig. 5.2 Variance density spectrum of the time series in Fig. 5.1

$$S_f = \left| 1 + 2 \sum_{k=1}^{k=m} D_k \, r_k \cos 2\pi f \, k \right| \tag{5.2}$$

This formula represents the Fourier transform of the autocorrelation function for the frequency being considered (for a step j, $f = j/2m$); r_k is the autocorrelation coefficient, D_k is a weighting function ('filter') that allows an unbiased estimate of S_f; k has the same meaning as in Eq. 5.1. The filter used is Tukey's [4], whose expression is: $D_k = (1 + \cos\pi/m)/2$.

5.3.2 Cross-Analysis

A cross-correlogram is calculated in the same way as a simple correlogram and with the same truncation value m. The difference is that the correlation is performed between two discretized time series, with the former $(x_1, x_2, ..., x_m)$ being considered the cause of the latter $(y_1, y_2, ..., y_m)$ (Fig. 5.3). The formulae are those proposed by Box and Jenkins [2]; since there is no symmetry in the correlations of x in y (designated $r_{x,y}$ or r_{+k}) and the correlations of y in x (designated $r_{y,x}$ or r_k), the calculation is undertaken from $-m$ to $+m$:

$$r_{+k} = r_{x,y}(k) = \frac{C_{x,y}\,k}{S_x.S_y}; \quad C_{x,y}\,k = \frac{1}{n}\sum_{1}^{n-k}(x_t - \bar{x})(y_{t+k} - \bar{y}) \tag{5.3}$$

$$r_{-k} = r_{x,y}(k) = \frac{C_{y,x}\,k}{S_s.S_y}; \quad C_{y,x}(k) = \frac{1}{n}\sum_{1}^{n-k}(y_t - \bar{y})(x_{t+k} - \bar{x}) \tag{5.4}$$

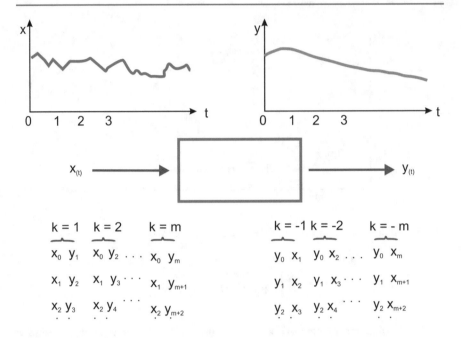

Fig. 5.3 Cross-analysis scheme

$$k = 0, 1, 2, \ldots, m \quad y \quad S^{2x} = \frac{1}{n}\sum_{1}^{n}(x_t - \bar{x})^2; \quad S^{2y} = \frac{1}{n}\sum_{1}^{n}(y_t - \bar{y})^2 \quad (5.5)$$

The asymmetry of the cross-correlogram (Fig. 5.4) for positive and negative k leads to the need to express the spectrum by a complex number: spectrum (f) = $k_{x,y}(f) - i\, q_{x,y}(f)$, where f is the frequency ($i = j/2m$) and i is the imaginary number. $k_{x,y}(f)$ is called *co-spectrum* and is expressed as follows:

$$k_{x,y}(f) = 2\left|r_{x,y}(0)\right| + \sum_{1}^{m}\left|r_x(k) + r_y(k)\right|D(k)\,\cos 2\pi f\,k \quad (5.6)$$

$k_{x,y}(f)$ is called the *quadrature spectrum*; it is obtained from the expression:

$$q_{x,y}(f) = 2\sum_{1}^{m}\left|r_x(k) + r_y(k)\right|D(k)\,\cos 2\pi f\,k \quad (5.7)$$

As in the simple spectrum, $D(k)$ is a filter that allows unbiased values to be obtained. For the cross-analysis, it is necessary to choose the same filter as in the simple analysis. Another possible way to write a spectrum expression is the complex form:

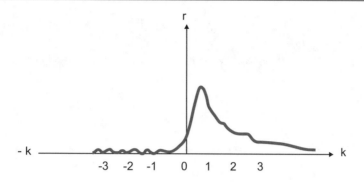

Fig. 5.4 Cross-correlogram

spectrum $(f) = |S_x, (f)| \exp[-i_{x,y} (f)]$, where $|S_{x,y}(f)| = \sqrt{k^2 x_{x,y}(f) + q^2_{x,y}(f)}$
is the *cross-amplitude function* (Fig. 5.5a), and $\theta_{x,y}(f) = \text{arc tag}\left[\frac{q_{x,y}(f)}{k_{x,y}(f)}\right]$ the *phase function* (Fig. 5.5b).

This second expression is used much more, given that $S_{x,y}(f)$ and $\theta_{x,y}(f)$ have physical significance and provide information that is relevant to linear systems. Since the simple spectrum can be interpreted as the value of the variance for each frequency and the cross-spectrum as the value of the covariance, it is interesting to define two other functions from them. The first $C_{x,y}(f) = \frac{S_{x,y}(f)}{\sqrt{S_x(f).S_y(f)}}$ corresponds to a correlation established for each frequency. This function is called *coherence* (Fig. 5.6a); it is similar to the cross-correlation coefficient of the time series x and y in the frequency domain. The second, $g_{x,y}(f) = \frac{S_{x,y}(f)}{S_x(f)}$, is a parameter similar to a regression coefficient of the x and y series, for each frequency. $g_{x,y}(f)$ is called *gain function* (Fig. 5.6b).

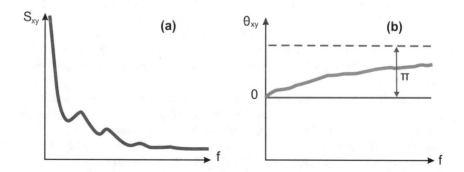

Fig. 5.5 a Cross-amplitude function; and **b** phase function

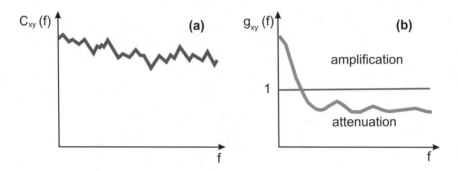

Fig. 5.6 a Coherence function; and **b** gain function

All these techniques are set out in specialized programs or sets of programs. One of them, for karst hydrogeology, is prepared by a team from the Diputación de Alicante. It can be requested from http://www.ciclohidrico.com, mfmejuto@diputacionalicante.es or Ciclo Hídrico, Diputación de Alicante, Avenida de Orihuela, 128. 03006-Alicante.

Two examples have been carefully studied and compared with three others analysed by Mangin in Chap. 4 [9, 10]. These are the springs of La Villa (Torcal de Antequera, Malaga), Fuente Mayor de Simat de Valldigna (Sierra Grossa Unit, Valencia) (see Box 1); and Aliou, Le Baget and Fontestorbes, the three last in the French Pyrenees (Fig. 5.7: see Box 2). The application has been extended to several outcrops in the province of Alicante (Box 3) and northeast Bulgaria (Box 4).

5.3.3 Using Correlation and Spectral Analysis

When a series of flows with a simple structure is the object of study, interpretation of the functions mentioned above presents no difficulty. It is enough to follow the standardized models and ranges of validity that can be obtained from the bibliography, for example when dealing with aquifers with a high regulatory power or, the opposite, with a very fast and immediate impulse response. However, interpretation —especially the function of cross-analysis in the domain of frequencies—becomes complicated for aquifers in which there are elements that are highly transmissive alongside others that are weakly karstified. As is well known, this is typical of most karst aquifers.

Another issue to consider in correlation and spectral analysis is the linearity of the input functions. Unless statistical testing is employed, the results obtained by manipulating data (system memory, duration of impulse response and concentration time) are clearly different for a series constituted of several hydrological cycles and for analysis of each individual hydrological cycle. Is this dissimilar behaviour due to the absence of linearity in the series or to the variation in the spatiotemporal characteristics of the aquifer?

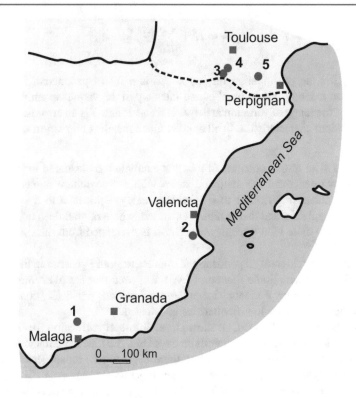

Fig. 5.7 Location of the karst systems studied: (1) Torcal de Antequera; (2) Simat de Valldigna; (3) Aliou; (4) Baget; (5) Fontestorbes

In what follows, we will provide guidance on the possible considerations in interpreting the results obtained from series of flow and precipitation data for karst springs, based on real examples that have been generated synthetically [5, 6].

5.3.4 Cross-Analysis

5.3.4.1 Synthetic Examples

To address the issue of interpretation of cross-analysis, we generated three series of synthetic flows from a known convolution nucleus to obtain the results of its application. We thus gained the advantage of knowing perfectly the type of aquifer that gives rise to the output function. As an input function, the daily rainfall from 1 September 1974 to 30 October 1981 (2192 data items) at the El Torcal meteorological station was used. We generated the output with a convolution function of the type:

$$S_t = E^t + \sum_{j=0}^{n} P_{t-j}\lambda_j \tag{5.8}$$

where P_{t-j} is the precipitation on day t_j; λ_j is a set of parameters forming the convolution nucleus, whose number and intensity in the various examples studied; in all cases the range of variation is between 0 and 1; and E is an error function with almost random characteristics. In all applied functions, the observation window was $m = 100$.

Example 1 The results obtained in this first example are illustrated in Fig. 5.8. In this case, we generated the synthetic series with a convolution nucleus of three parameters decreasing from 1 to 0.1, without delay, equivalent to a system with barely any regulation and that is highly karstified, where rainfall does not remain in the aquifer for more than three days and most is evacuated from the spring on the first day.

The cross-correlogram responds to the impulse response generating the exit flow data. In the cross-amplitude function (FAC), it is seen that the filter introduced by the input signal for any frequency is not large and remains practically constant; only a small increase at very low frequencies is detected.

The coherence function (FCO) shows a small loss of correlation for frequencies above 0.33 for periods of less than three days (Fig. 5.9), which coincides with the regulation introduced at the point when there begins to be an appreciable attenuation of the signal, as evidenced by the gain function (FGA).

Despite not having introduced any delay, the phase function (FFA) shows good alignment at frequencies lower than 0.33, from which there is enough distortion attributable, as we have seen, to the loss of signal in the high frequencies (Fig. 5.10). Although it appears that no delay has been introduced, there is indeed an observable difference, and it coincides with the time when the 'weight' of the convolution nucleus has become 50% of the total weight of the parameters; that is,

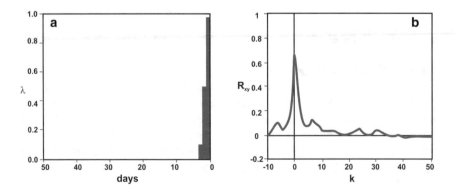

Fig. 5.8 **a** Convolution core parameters; and **b** cross-correlogram

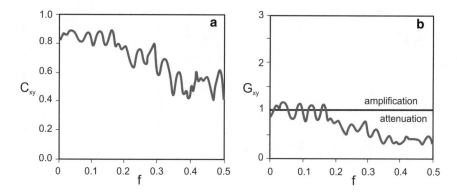

Fig. 5.9 a Consistency function; and **b** gain function

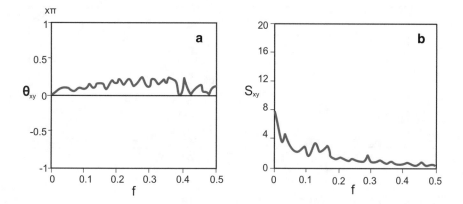

Fig. 5.10 a Phase function; and **b** cross-amplitude function

it is equivalent to the centre of gravity of the unit hydrograph. In the FFA, the offset is given by $d = /2\phi\pi f$, where ϕ is the value of the function and f the frequency; therefore, the delay is directly proportional to the angle of the line $\phi = 2\pi df$, which is worth $2\pi d$. Calculation of the FFA results in 0.25 days.

Example 2 In this second case, we have a synthetic generation of an output register from a strongly regulating convolution nucleus, made up of 50 parameters decreasing from $\lambda_0 = 0.4$ (Fig. 5.11). The cross-correlogram has a very loose shape, characteristic of a highly regulating system in which the concentration time is barely apparent.

There are a strong filter and great signal attenuation in the high frequencies, favouring the low ones that present very high values; this is observable in the FAC and FGA (Fig. 5.12). However, this decrease is not abrupt, but is produced by a gradual decrease between the frequencies of 0.04 and 0.2 and shows practically null values for higher frequencies.

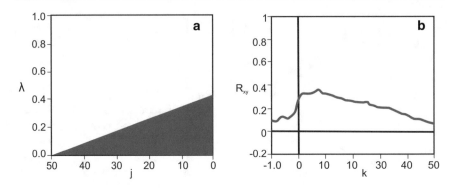

Fig. 5.11 **a** Convolution core parameters; and **b** Cross-correlogram

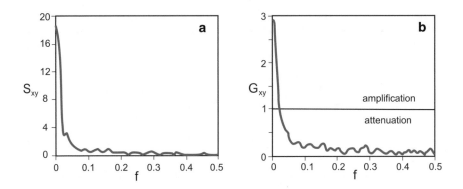

Fig. 5.12 **a** Cross-amplitude function; and **b** gain function

This can also be discovered in the FCO, showing that variations in precipitation between these two periods, 25 and 5 days, still have a response in the output signal, although much filtered and attenuated (Fig. 5.13). The FFA shows alignment only at frequencies between 0 and 0.02, above which the signal is so attenuated that distortions occur in the FFA values. This alignment shows a delay of 14 days, which corresponds to the position of the 'centre of gravity' of the convolution core generator of the synthetic series; therefore, it is equivalent to the response time of the system.

Example 3 Finally, we generated an output signal from a convolution core made up of 25 parameters, an initial one high ($\lambda_0 = 1$) followed by 24 with the maximum located at $\lambda_5 = 0.3$. It represents an impulse response that is a combination of the two previous cases, in line with the actual operation of most karst aquifers: an immediate response characteristic of the most transmissive part of the aquifer; and another that shows a certain modulation of the input signal, characteristic of the most regulating part of the aquifer. The cross-correlogram clearly shows the two

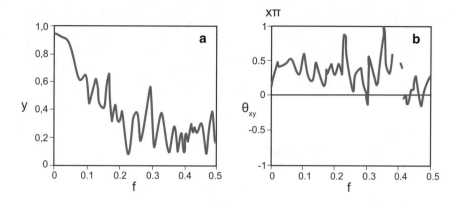

Fig. 5.13 **a** Coherence function; and **b** phase function

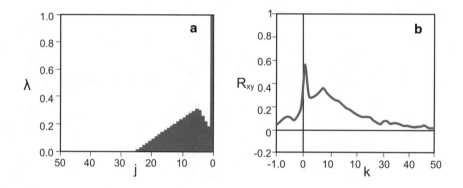

Fig. 5.14 **a** Convolution core parameters; and **b** cross-correlogram, Example 3

peaks typical of a system with two flow components, one initial without delay and the other located at Step 5. It is identical to the impulse response (Figs. 5.14 and 5.15).

The FAC and FGA functions (Fig. 5.16) show a strong filter and attenuation, respectively, of the input signal for frequencies higher than 0.04, in favour of the low frequencies. However, and this is important when interpreting real cases, it cannot be said that it is annulled, like in the previous example. Instead, it is maintained around values of one for the FAC and around 0.4 in the FGA, as in the first example where there was scarcely any regulation. Therefore, we can affirm that, although in the rest of the cross-functions the presence of a rapid component in the aquifer is undetectable, its footprint is perfectly marked in the FAC and in the FGA.

If we assume that FGA values above one (corresponding to output signal gain) constitute the aquifer buffer fraction, and that values below 0.4 are considered null,

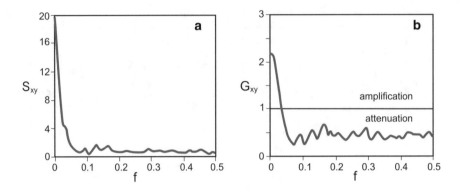

Fig. 5.15 **a** Cross-amplitude function; and **b** gain function from Example 3

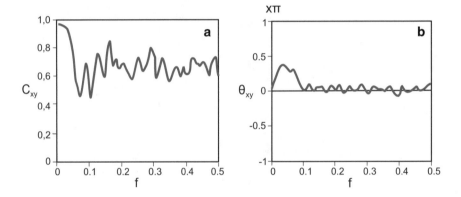

Fig. 5.16 **a** Coherence function; and **b** Phase function from Example 3

then the frequency that coincides with the value of one will correspond to the duration of the buffer part of the aquifer; the frequency at which FGA begins to take values below 0.4 will correspond to the duration of the most transmissive part. According to this criterion, in this example the frequency corresponding to FGA = 1 is around 0.4, a period of 25 days, approximately equal to the duration of the regulating part of the introduced impulse response. The FGA does not have values lower than 0.4 in any frequency range; therefore, the duration of the most transmissive fraction must be less than two days (frequencies higher than 0.5). Remember that its duration, imposed on the convolution nucleus, is one day.

The correlation observed in the FCO in the mid- and high frequencies is even higher than in the first example, around 0.7. The delay, calculable in the low frequencies of the FFA, is seven days, which corresponds to the regulatory part; for frequencies higher than 0.1 the delay is zero and corresponds to the fast circulation part [5].

5.3.5 Applications

We carried out the cross-analysis with a 100-day observation window of the precipitation and flow at four karst springs: El Torcal and Simat de Valldigna; and two located in the French Pyrenees, Aliou and Baget, whose characteristics have been already described.

The cross-correlogram in Fig. 5.17 clearly shows this difference in response to precipitation at the four springs. On the one hand, the duration of the response in Aliou and Baget is very short, of the order of 12 and 20 days, respectively. By contrast, in Simat it fluctuates between 60 and 80 days, and in El Torcal, it is longer than 100 days. On the other hand, in Aliou and Baget the response is immediate, whereas in Simat the concentration time is two days and in El Torcal it fluctuates between 12 and 35 days.

In both Aliou and Baget, there is a series of spikes in the cross-correlogram after the initial one. This could be regarded as being caused by the presence of another flow component within the aquifer itself; however, these peaks are reflected in the negative part of the abscissa axis, resulting in a cross-correlogram with symmetry, apart from the initial sharp decrease.

The answer to this behaviour is to be found in the low modulation that the system exerts on the input function, so that the periodic components of precipitation are detected in the flows. In the case of precipitation in the Pyrenees, there is a notable periodicity of 30 days in Chap. 4 [13].

Figure 5.18 shows how in El Torcal the FAC is practically annulled for frequencies higher than 0.05 (periods lower than 20 days). In the other three aquifers, although there is a clear decrease in the function in the mid- and high frequencies in favour of the low ones, this is not completely annulled in Aliou and Baget, and in

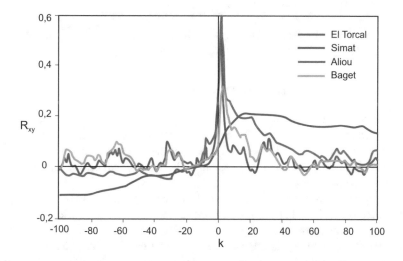

Fig. 5.17 Cross-correlograms of the Pyrenean and Mediterranean aquifers

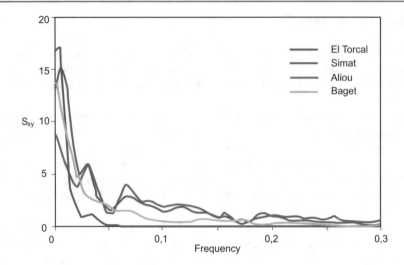

Fig. 5.18 Cross-amplitude functions of the series of studied aquifers

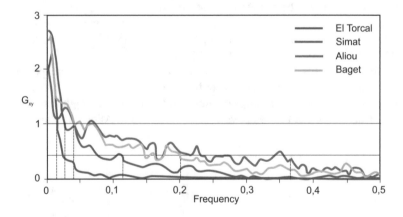

Fig. 5.19 Gain functions of the series of aquifers studied

Simat, it cannot be considered null except for frequencies higher than 0 to 150.20. In other words, as in Example 3, there is clearly a rapid circulation in these last three aquifers that is absent from El Torcal, but there is also a 'regulated' fraction; this term is subjective, since it will depend on how the length of time between entry and exit is considered.

This quantification can be obtained more easily from the FGA in Fig. 5.19. Indeed, let us consider that the value of one of the FGA coincides with the duration of the impulse response of the regulatory fraction of the aquifer, and that the value of 0.4 corresponds to the duration of the rapid fraction. Between these two ranges, we have intermediate flow conditions. These values coincide with those obtained from the synthetic examples.

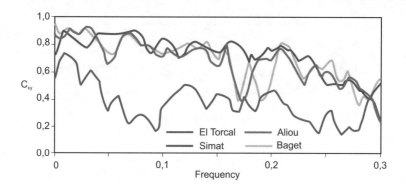

Fig. 5.20 Coherence functions of the series of aquifers studied

Transforming frequencies into periods, under these conditions we will find that the duration of the regulatory response in El Torcal is 63 days, in Simat 38 days and in Aliou and Baget 24 days. The duration of the rapid response is 38 days in El Torcal (in this case, it cannot be considered rapid at all), nine days in Simat, five days in Baget and three days in Aliou. On the other hand, the fact that the duration of the rapid response in Aliou is three days does not mean that the spring does not respond immediately to rain with a considerable flow but that for three days the water circulates essentially through highly transmissive conduits. This separation criterion can be modified for various cut-off values in the FGA.

The FCO in Fig. 5.20 confirms what we have found so far. The case at El Torcal is identical to the second synthetic example. There is a major decrease in coherence for periods of less than 60–50 days. In the other three aquifers, even in Simat where the regulatory fraction of the system is quite important, it is maintained with relatively high values for frequencies lower than 0.25 (periods of longer than 5 days), indicating the presence of a rapid circulation component.

The FFA study (Fig. 5.21) makes it possible to calculate the delay in frequencies after the first points and before the signal is distorted (corresponding to rapid circulation in the system), in a similar way to the procedure in the third synthetic example. Thus, for Simat the delay is one day, while for Aliou and Baget, it is half a day. Strange as it seems, delays of less than a day can be calculated even though the entry and exit functions are discretized at daily intervals, as the delay corresponds to the 'centre of gravity' of the response.

5.3.6   Linearity

Analysis of a simple correlogram can provide answers to some questions about this series of springs. A comparative study was carried out to observe in simple correlograms the variation in the flow of the individual cycles against the overall series, as shown in Fig. 5.22. This confirmed that there is a notable dispersion in the correlograms, especially for the spring at Fuente Mayor.

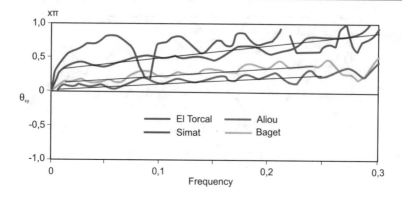

Fig. 5.21 Phase functions of the series of studied aquifers

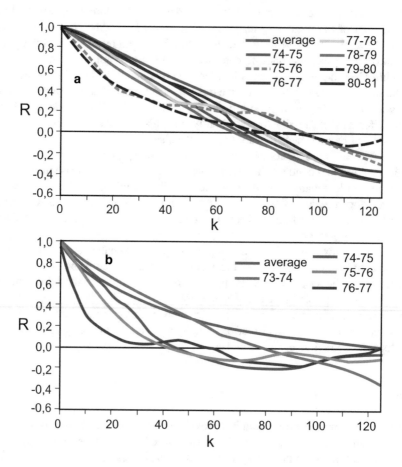

Fig. 5.22 Simple correlograms of each of the hydrological cycles: **a** La Villa; and **b** Fuente Mayor

Adopting the criterion of regarding as null any values in the simple correlogram below 0.2, the memory effect in La Villa was found to fluctuate between 45 and 75 days, and in Fuente Mayor between 15 and 57 days. The dispersion observed may be due mainly to the lack of linearity in the systems. The behaviour of the aquifer depends on its state at the time of rainfall, especially regarding the level of saturation and the percentage of infiltration. To see to what extent the absence of linearity is the cause of the distortion in the form of the correlograms, two synthetic series of flows were simulated, generated by applying a linear convolution function to the rainfall data of El Torcal and Simat. For El Torcal, the chosen convolution core has a total of 100 parameters and for Simat 65, both decreasing (Fig. 5.23). Figure 5.24 shows the correlograms obtained for each of the cycles, as well as the mean correlogram.

From the results, it is clear that the criterion of taking the value of the correlation coefficient of 0.2 as the basis for estimating the memory effect is quite inaccurate. The memory effect imposed on El Torcal and Simat is 100 and 65 days; however, by using this criterion, an average value of 75 days was obtained for El Torcal, with a variation of between 30 and 77 days, and for Simat an average value of 34 days, with a range between 27 and 37 days. This is a consequence of the influence of the seasonal correlation, which in these cases acts in favour of a reduction of the system's own dependency.

On the other hand, it was confirmed that the correlograms of the synthetic series of flows of El Torcal present more variation than the series of real flows. This is difficult to attribute either to the state of the aquifer or to the percentage infiltrated, since the aquifer itself has no influence over the generation of synthetic flow.

Not even the amount of precipitation in the hydrological cycle has much influence, as the fastest decrease in the correlogram is observed in Cycle 80–81. Although the driest is Cycle 74–75, at 528 mm, the correlogram does not reveal any visible characteristic to differentiate it from the wettest, Cycles 76–77 and 78–79, both at 1047 mm. Therefore, the lack of linearity in the system, as we showed, is not the main cause of the variation observed in the correlograms of the flows in the annual cycles, and it is necessary to look at in two causes:

- The lack of stationarity in dealing with series with a sampling interval of a single hydrological cycle. The hypothesis of stationarity that is assumed in principle for a series of flows of several cycles of length is questionable when dealing with a small series, especially in highly regulated systems. As shown by the spectral density function of flows in Chap. 4 [12, 13], there is a marked annual periodicity. Therefore, the presence of important tendencies within each cycle, causing the second-order stationarity hypothesis to be rejected, should not be discounted.
- A second cause is due, as Mangin in Chap. 4 [9] observed, to the form and distribution in which rainfall occurs. Very regular rainfall on an aquifer with a linear response would give correlograms with less pronounced angles than irregularly distributed rainfall, regardless of the amount falling and with little influence from the state of soil saturation.

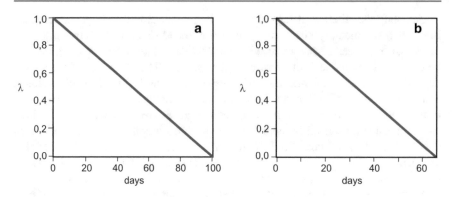

Fig. 5.23 Convolution nuclei used in the generation of synthetic flows: **a** El Torcal and **b** Simat

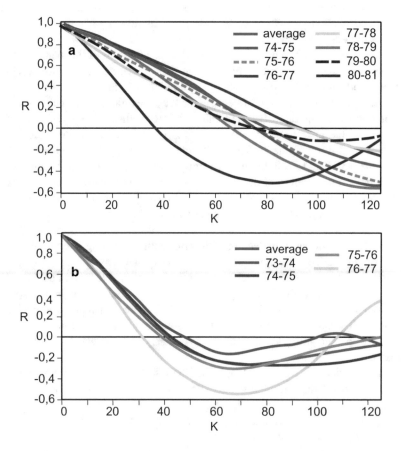

Fig. 5.24 Simple correlograms of the hydrological cycles of the synthetically generated flow series with rainfall data: **a** El Torcal and **b** Simat

5.4 Box 1: Worked Examples of Cross-Analysis

5.4.1 Torcal de Antequera

Torcal de Antequera is an area of 28 km^2 made up of carbonate rocks. Its highest point is Chamorro Alto (1369 m asl), and the average altitude of the massif is approximately 1000 m. The aquifer material is composed of limestone and dolomite at the base, of Jurassic age, with a thickness of greater than 600 m. The structure of the Torcal is an anticline with strong development of the trough sector. The whole is intensely fractured.

The massif is drained essentially through the spring of La Villa (elevation 586 m), whose discharge has been controlled since October 1974. Since the spring is used as the water supply for the city of Antequera, not all the discharge is measured at the gauging station and has had to be estimated for the period of observation. In addition, there is a borehole in the vicinity of the spring to ensure an urban water supply, and this has affected the natural discharge regime. The maximum discharge measured at the spring was 1785 l/s in February 1979, and the minimum has been estimated at 90 l/s, for the whole period considered.

There is a rainfall gauge at El Torcal, and its data are representative across the massif. However, to establish an identical period of observation of discharges and rainfall, data for the period prior to January 1975 were obtained from Antequera, 4 km north of the aquifer.

5.4.2 Sierra Grossa

The Fuente Mayor de Simat de Valldigna is one of the mainsprings of the major Sierra Grossa aquifer unit, with an area of approximately 370 km^2. The spring is on the north-eastern edge and, although its catchment is not known in great detail, it is estimated to drain around 20 km^2. The top of its catchment is Mondúber peak (841 m), and its average altitude is around 400 m.

The aquifer materials are made up of the Creu and Jaraco in Chap. 2 [15] formations, essentially limestone and dolomite in this sector, with some intercalations of sand, silica gravel and yellowish marl in the former, and limestone and dolomite with marly intercalations in the latter. Near the spring is the polje of Barx-La Drova, with a catchment area of 8 km^2. Within it are several ponors, the most important being La Doncella. These quickly absorb the run-off into the polje during heavy rainfall. The discharge of the spring began to be controlled in 1973. The maximum flow measured was 957 l/s and the minimum 80 l/s. The rainfall gauging station, considered to be representative of the basin, is Simat de Valldigna.

The other karst systems analysed, which will be compared with these two examples, are those of Aliou, Le Baget and Fontestorbes. Aliou, with an area of 11.93 km^2, has an average discharge of 475 l/s, with extreme flows approaching 20 m^3/s and falling to less than 10 l/s. Le Baget is a system made up of karstified limestone and practically impervious materials extending across 13.25 km^2, and its average discharge is 530 l/s. Finally, Fontestorbes, a complex karst system that is partly confined, occupies an area of 85 km^2 and has an average annual discharge of 2350 l/s.

5.4.3 Procedures Undertaken

The discharge of each spring has been considered as the output of a system, assumed to be linear assumption, where rain corresponds to the input. In other words, there is a system with an *input function* from which an *output function* is obtained. If these two functions are known, we can obtain information on the behaviour of the system, which depends essentially on its structure and its operation. In the cases studied, the average daily discharges constitute the outflow function of the aquifers mentioned. Since the treated series are not stationary, strictly speaking it is important to choose an appropriate time, since the results are dependent on this choice. It is noted that the best results are obtained at end of a low-water level and the start of a new hydrological cycle.

5.4.3.1 Simple Analysis

Rainfall
Figure 5.25 shows the correlograms of rainfall at the Torcal and Simat de Valldigna rain gauge for the periods between 5 February 1975 and 30 September 1981 (i.e. a little more than 6 years) and 18 October 1973 and 30 September 1981 (8 hydrological years). The average rainfall was 2.3 and 1.9 mm/day, respectively, and the variances were 71.3 and 93.7.

For the Torcal station, the shape of the correlogram for the short-term analysis ($m = 125$ days) shows that rainfall could be considered as a *pure random function*; however, in the corresponding spectrum, there is a trend of very low frequencies and a monthly cycle (Fig. 5.26).

With a wider *observation window* ($m = 1250$ days), the correlogram (Fig. 5.27) retains the random character of the input function (total absence of correlation between values), yet the corresponding spectrum shows an annual cycle of rainfall that determined the trend in previous analyses. This annual cycle is a known fact, at our latitudes (Fig. 5.28).

All that was said for Torcal, with slight variations is applicable also to Simat's rainfall. In effect, the annual cycle is less marked in the long-term and the assumed monthly cycle in Torcal is not seen in Simat. It may be thought

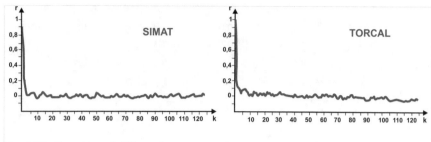

Fig. 5.25 Short-term rainfall correlograms

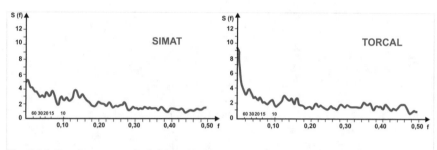

Fig. 5.26 Density spectra of rainfall variance (short duration)

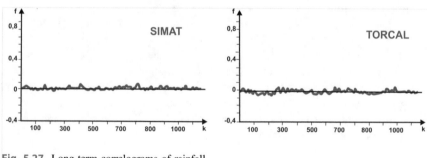

Fig. 5.27 Long-term correlograms of rainfall

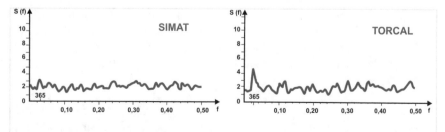

Fig. 5.28 Density spectra of rainfall variance (long duration)

that the discontinuity in rainfall in Simat is less marked than in Torcal; however, in both cases, at least in the study of short-term rainfall, the input functions can be assimilated to quasi-random functions.

Flow Rates Figure 5.29 shows the correlograms of the measured discharge of the springs of La Villa and Fuente Mayor, for an observation window of 125 days (the averages and variances for the period considered are, respectively, 424 and 198 l/s, and 80 and 1 l/s). In the case of the La Villa spring, it can be deduced that the Torcal aquifer acts as a system with great inertia, with an important memory effect. This indicates that the system has considerable reserves. Its regulation time is long (73 days). Given that there is a very high degree of correlation between these events, the fact that the input function influences the output function only for long-term events greatly facilitates the modelling of these systems. Traditional models can be used, which require few parameters to be adjusted.

The spectrum (Fig. 5.30) shows that the system lets practically no variations in rainfall to be seen in the flow; this does not hold for the low-frequency domain (i.e. long periods). With regard to the Fuente Mayor, a certain short-term memory effect is observed, and this indicates the presence of a discontinuity in the behaviour of the aquifer. A comparison of the rain and discharge spectra shows the presence of visible rain components in the flows, for the medium frequencies and especially for the high ones (for short periods). From 4.5 days, these components are totally filtered.

This discontinuity observed in the correlogram could be due to an anomaly of operation caused by the Barx polje, which in the flood period acts as a collector and makes a sudden contribution of water to the system, ensuring the transfer of a large part of the details of the rainfall events. After several days, the system shows a more inert behaviour. One might also think that there is a higher level that is more karstified, and that this works exclusively at times of high waters and with three-phase flow. On the other hand, the regulation time of this aquifer, although long, is less than that of Torcal (52 days, in this case).

Using a wider observation window (1250 days), annual cycles are observed in the two correlograms (Fig. 5.31) and are much more pronounced in the spring of La Villa than in the Fuente Mayor. From the observation of the long duration spectrum (Fig. 5.32), it can be deduced that, in addition to the annual cycle, there is a seasonal component of approximately three months, which is also more pronounced at La Villa. This seasonal component is observed in the rainfall. From this, it can be deduced that its origin is climatic and thus reflects the effect of evapotranspiration.

Although no seasonal component is detected in the rainfall, a very strong seasonal trend (low frequencies) is observed in the flow of Fuente Mayor. This could be attributed to the degradation of its reserves over time. It is

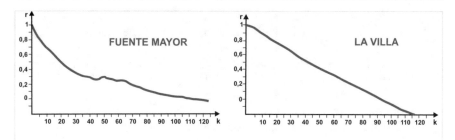

Fig. 5.29 Short-term correlograms of the discharge of the Fuente Mayor (Simat) and La Villa springs (Torcal)

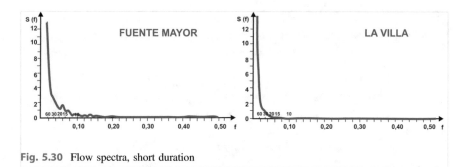

Fig. 5.30 Flow spectra, short duration

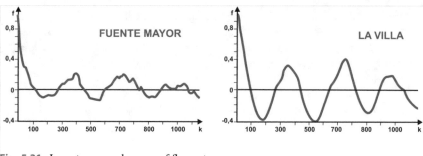

Fig. 5.31 Long-term correlograms of flow rates

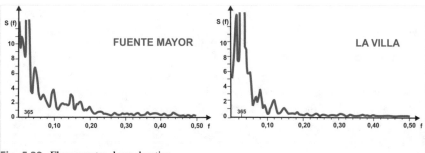

Fig. 5.32 Flow spectra, long duration

likely that this trend could be the result of pumping from boreholes in the vicinity of the spring.

This observation leads to another conclusion: the inertia of the system is not due to the part of the aquifer situated downstream of the polje but to that situated upstream, since when the inputs are high the transfer of water is rapid. There is therefore a fairly karstified aquifer downstream, not upstream, which reflects a certain complexity, the result of an evolution that is also complex.

5.4.3.2   Cross-Analysis

Cross-correlation makes it possible to establish the relationship between cause and effect, both in time and in frequency. The cross-correlogram provides a fairly satisfactory approximate representation of the *impulse response* of the system, a concept equivalent to the *unit hydrograph* usually used in surface hydrology or system response to a unit input. In the negative part of the axes, the correlogram should be null. Any deviation from zero indicates the possible influence of measurement errors responsible for background noise.

The cross-correlograms obtained are shown in Fig. 5.33. For El Torcal, the impulse response of the system has a very flattened profile. The correlogram also shows notable background noise for those values corresponding to the end of the impulse response. This indicates that the estimated low-water flow has a considerable error, as previously indicated. The duration of the impulse response of the system is very long, indicative of its great regulatory power. One might regard this behaviour as very similar to that of a porous medium.

In contrast, the cross-correlogram of the Fuente Mayor shows a more pointed and less extended impulse response. The initial sharp peak is indicative of the system's more pronounced response to rainfall at first (fast flow) and is interpreted as an effect of the influence of massive water inputs to the polje drained by the ponors. Later, slower recession would be more similar to that of a porous media.

The cross-amplitude functions obtained are shown in Fig. 5.34. In the case of Torcal, it is observed that most of the details of the rainfall events are filtered by the system, allowing only a small part to pass in the low-frequency domain (long period). In the case of the system linked to the Fuente Mayor, the details are much less filtered; the flows in this case translate in a more pronounced way to variations in the input function.

The phase function allows the phase difference between input and output to be calculated. Using the expression $T = \theta/2\pi f$, we can quantify this phase difference (T) in days, depending on the frequency (f) and the phase (θ). For a frequency of 0.10, the delay is 2.3 days for Fuente Mayor, and 1.8 days for El Torcal (Fig. 5.35).

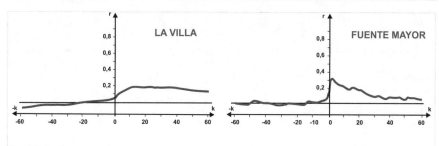

Fig. 5.33 Cross-correlograms

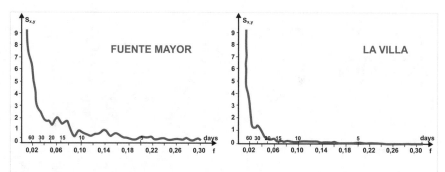

Fig. 5.34 Cross-amplitude functions

The coherence function allows us to visualize the degree of interrelation between rainfall and flow in the frequency domain. It also allows us to control for previous data procedures and provides information on the linearity of the system. For the Torcal aquifer, coherence is very low or even nil in the medium- and high-frequency domain, while for the Simat aquifer coherence is acceptable in the medium and low frequencies but less so in the high-frequency domain. That is why the phase function does not have any obvious meaning, in the case of the Torcal aquifer (Fig. 5.36).

With respect to the gain function, as with the functions already analysed, the difference between Torcal and Fuente Mayor is pronounced (Fig. 5.37). In fact, although in the two functions there is a very strong gain in the domain of low frequencies (long periods), to the detriment of the medium and high ones, for El Torcal the drop is very sharp, with an almost total attenuation of 0.06. By contrast, in the Fuente Mayor the amplification is more important and the attenuation is progressive, without sudden jumps.

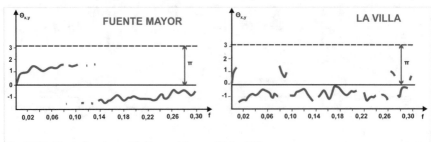

Fig. 5.35   Phase functions: (1) Fuente Mayor; (2) La Villa

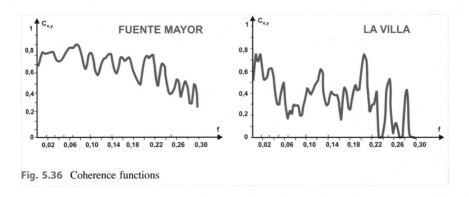

Fig. 5.36   Coherence functions

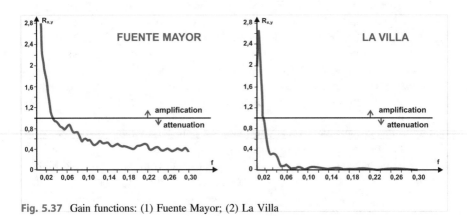

Fig. 5.37   Gain functions: (1) Fuente Mayor; (2) La Villa

5.5   Box 2: Comparison to Systems in the French Pyrenees

Annual precipitation cycles in the Pyrenees are practically impercep-
tible, unlike at Torcal and Simat. The same is true of shorter cycles,
such as the seasons. Simple rain correlograms decline rapidly and
become null and void in less than five days. The spectra, although
monotonous, show a series of peaks with values close to the background yet which have a certain importance, as will be indicated later.
Therefore, as far as the distribution of rainfall in time is concerned,
there is a notable difference between the climate of southern Spain and
that of the Pyrenees.

Figure 5.38, left, shows the correlograms of the three springs of Aliou, Le
Baget and Fontestorbes, corresponding to the available time series of 11, 12
and 14 years, respectively. The analysis was also carried out for a truncation
value of 125 days. It can be observed that the values decrease rapidly in
Aliou and Baget, whereas in Fontestorbes the correlogram has a much slower
change in form. Aliou's correlogram has many features in common with the
rainfall correlogram, while the Fontestorbes correlogram does not record it at
all, and Baget's correlogram is intermediate. All of this is equally visible in
the spectrum. The *regulation times* are 14 days for Aliou, 22.5 days for Baget
and 50 days for Fontestorbes. From this, it can be concluded that the Aliou
system lacks reserves and the Fontestorbes system has considerable regulatory power. In the latter, its greater extent and complexity must play a relevant role (Fig. 5.39, right). Figure 5.40 shows simple correlograms for
Torcal, Simat, Sagra and Baget.

In the three spectra, there are peaks for the same frequencies that are also
found in the rainfall at the same sites, enlarged in Aliou and attenuated in
Fontestorbes. These peaks correspond, then, to the distribution of the rainy
periods. Details of the rainfall events, for Aliou's system, are transmitted

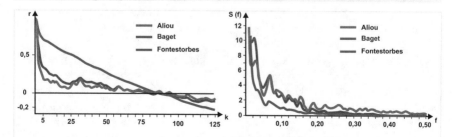

Fig. 5.38 Correlograms, left, and variance density spectra, right

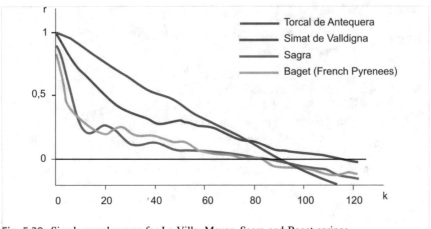

Fig. 5.39 Simple correlograms for La Villa, Mayor, Sagra and Baget springs

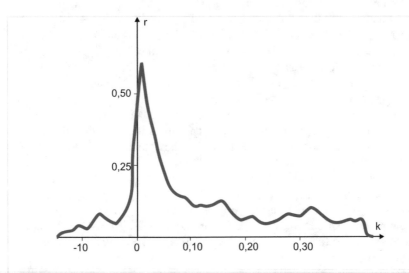

Fig. 5.40 Baget's cross-correlogram

almost completely unmodified, while in the other two systems they undergo an important filtering.

Comparison of these data with those obtained for the Spanish massifs described here reveals marked differences, interpreted as due to differences in structure and, above all, in operation. As opposed to a regulation time of 14 days in Aliou, it is 73 days at El Torcal, without taking into account the differences already pointed out in the entry signal (rainfall). Aliou and Torcal represent the opposite extremes of the examples treated; while one transmits the variations in rainfall pass almost unchanged through its system, the other lets changes pass only at very low frequencies (long periods); in Aliou, reserves are practically absent, and in Torcal, by contrast contrary, they are considerable. From a practical point of view, this means that the chances of success in a borehole survey are very high in Torcal and practically nil in Aliou. The regulation time of Fontestorbes (85 km^2) is equivalent to that of Simat (20 km^2), 50 days, which shows a certain similarity between the two, although the surface of the system can play an important role.

The long-term analysis ($m = 1800$ days in these cases) also shows an annual cycle in the flow, very marked in Fontestorbes and slight in Aliou; as this cycle is not visible in the rainfall here, it is interpreted as having been induced by the system itself, with a memory effect the stronger, the larger the reserves. One might regard the presence of this annual cycle as attributable to evapotranspiration.

When comparing the cross-correlogram of Baget (Fig. 5.41) to those obtained for Torcal and Simat (Fig. 5.33), a great difference emerges between them. While Torcal's has a flattened profile, Baget's has a marked peak, and Simat's is intermediate between the two. They are therefore very different systems, which is of great practical importance in terms of their behaviour and how they are studied. In the first case, the transit of water is very fast, and in the others, it is much slower.

The existence of several decreasing amplitude modes in the unit hydrograph reflects, on the one hand, the non-stationary nature of the time series and, on the other hand, the imperfect linearity of the system; due to this, the impulse response varies, depending on the initial state of the system and the shape of the input impulse. The different operation of the compared systems is also manifested in the phase function (delays of 0.8, 0.1 and 1.5 days for

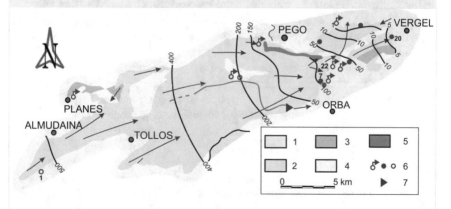

Fig. 5.41 Hydrogeological scheme of the Alfaro-Mediodía-Segaria unit (Pulido-Bosch 1977): (1) Detrital materials; (2) Karst materials; (3) Oliva formation—lower Cretaceous; (4) Belgida formation; (5) Valencia group, Keuper; (6) Spring; (7) Isbert dam on the River Girona

Aliou, Baget and Fontestorbes, respectively). These data are essential to ascertaining the sampling rate when studying flood propagation in the system.

The coherence function of these three systems is quite homogeneous, although a series of thresholds are observed for various periods, interpreted as corresponding to changes in the state of the system in relation to the rainfall input in Chap. 4 [10]. In all there is a deterioration in coherence as frequency increases (lower coherence when the time period decreases). Of all the examples examined, it is undoubtedly Torcal's that presents the least coherence between rainfall and flow; due to this, the phase function does not admit an easy interpretation.

As regards the gain function, Aliou's system, devoid of reserves, alters the input signal very little; as systems' reserves increase (in increasing order, Baget, Fontestorbes, Fuente Mayor and Torcal), there is a marked attenuation of the input signal at high frequencies (low periods) favour of low frequencies (seasonal and annual); this is why the gain function gives a good insight into a system's regulatory effect.

In the examples considered, the analysis makes it possible to quantify aspects related to the structure and operation of the systems. As far as rainfall is concerned, the regime is different in southern Spain and the French Pyrenees; however, in both the rainfall can be considered as a quasi-random function (white noise). No marked cycles are detected in the French Pyrenees, while in Simat and Torcal annual and seasonal cycles are obvious. The rainfall regimes are also different in Torcal and Simat. For example, while 1979 was the driest year in Simat, in Torcal the rainfall was above average.

Comparing the spectra and correlograms of the discharge to those of the rainfall allows us to determine the system's degree of modification of details of rainfall events. We can see that, compared to karst systems such as Aliou that transmit all rainfall variations, there are others, such as Torcal, that reflect only long-term influences (annual cycle). This points to the importance of the regulatory power of the system, practically nil in the first and very pronounced in the second. Another parameter that can be quantified by this method is regulatory time is also, and 73 and 50 days are obtained for Torcal and Simat, respectively. These data are relevant to modelling the system and to interpreting the chemical characteristics of the water and the karstification.

Since the input function can be considered as a random function, the cross-correlogram is a good approximation of the system's impulse response (unit hydrograph) and signature. According to the structure and operation of the system, there will be various impulse responses. A very sharp impulse response of short duration (Aliou's case) points to the presence of a hierarchical karst network, with similar characteristics to a surface watercourse with a rapid concentration time. On the other hand, a flat impulse response of very long duration (Torcal's case) indicates a relatively homogeneous medium with great regulatory power. The combination of observations in the Simat

Table 5.1 Summary of significant parameters of four highly significant karst systems

Types	'Memory effect' ($r = 0.1-0.2$)	Spectrum (frequency)	Regulation time	Unitary hydrogram
Aliou	Very small (5 days)	Very large (0.30)	10–15 days	
Baget	Small (10–15 days)	Large (0.20)	20–30 days	
Fontestorbes	Big (50–60 days)	Narrow (0.10)	50 days	
Torcal	Considerable (70 days)	Very narrow (0.05)	70 days	

case signifies the coexistence of two behaviours, the first predominating in floods and the second in base flow.

The cross-amplitude, phase, coherence and gain functions provide identical information. The phase function also makes it possible to quantify the difference between the impulse due to rainfall and the variation in the resulting flow. This relates to the operation of the system.

These methods also provide conceptual information (Table 5.1). Indeed, it can be deduced from these results that there is no single model of karst aquifers. Instead, there is a wide range of models and the cases analysed represent two extremes (Aliou and Torcal), with intermediate types. The Aliou model, typified by a highly developed karst network with no significant storage in conduits or a rock matrix, transmits rainfall variations without much modification. The Torcal model, with its capacious storage and considerable reserves, filters the rainfall signal. This system is easy to model using 'traditional' methods. Moreover, the chance of success when sinking a borehole here is high.

5.6 Box 3: The Alfaro-Mediodía-Segaria Unit

The Alfaro-Mediodía-Segaria unit occupies 200 km^2, of which 175 km^2 is of permeable material (Fig. 5.42). With a length of 35 km and a maximum width of 10 km, it is essentially made up of limestone and dolomite from the Creu formation (Middle and Upper Cretaceous),

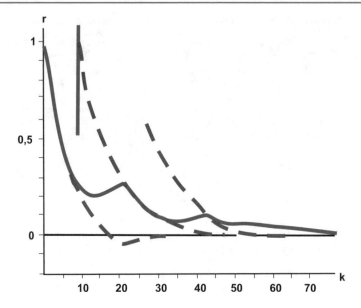

Fig. 5.42 Correlogram for short duration discharge; a dashed line shows a possible interpretation

which extends over 150 km^2, and limestone and calcarenite from the south (Eocene) and the Beniganim (Miocene) formations.

The western and northern borders are impervious, made from a thick marly series of the Belgida formation (Miocene). These materials also make up part of the eastern and southern borders, together with clay of the Tollos formation (Oligocene). The Sierra de Segaria and the eastern end of the Sierra del Mediodía are in contact with detrital materials in the aquifer of the Gandía-Denia plain. It is precisely along these edges that there is discharge of the unit, both visibly, through several springs, and in a hidden way.

There are many peculiarities about this unit's hydrogeological operation. The Isbert dam, with a capacity of approximately a hm^3, was built across the Girona valley, which transects the unit, to regulate the run-off of some 40 km^2 of catchment area. Since the dam is over karst carbonate materials, any water susceptible to being dammed infiltrates the aquifer. This recharge and the direct infiltration of rainwater constitute the two main feeds into the hydrogeological unit.

Discharge takes place through a series of springs, through pumping into constructed intakes and, in a hidden way, towards the detrital material of the Gandía-Denia plain. The springs are in various places and heights. One group is on the edge of the Pego marshlands (Balsa Sineu: Photo 5.1) very close to

Photo 5.1   Sineu marshland in May 2005 (Photo J. Ballesteros)

sea level, with water of sodium chloride type indicating a possible current or past marine influence.

The other group of springs, possibly *trop plein* of those already described, is in the vicinity of Sagra. At least three are known, and the one at the lowest elevation is actually a gallery—La Cava (Photo 5.2)—about a kilometre long, excavated at the beginning of the last century. At a higher level and in the Sagra village itself, there are three small springs that dry up in a pronounced drought. Finally, there is the spring of Tormos, or Bolata, the *trop plein* of this system (Photos 5.3 and 5.4), of variable discharge and very sensitive to heavy rainfall. It stops flowing in dry periods.

The Instituto Geológico y Minero de España (IGME) began to monitor the daily discharge of these springs by means of two limnometric scales, one in the Cava spring and the other in the Barranco de Bolata, obtaining fairly precise measurements of the discharge of the sector. There are now 84 months of data. The nearest available complete rainfall data are from Denia rain station, and this is relatively distant from the study area.

Figure 5.43 shows the short-term flow correlogram [7]. The memory effect obtained is of the order of 25 days, although one can interpret the presence of three impulses spaced over time: one very fast, typically karst, of 10 days of

Photo 5.2 Manantial de la Cava, April 2011 (Photo J. Ballesteros)

memory effect; another slower, of about 25 days' range; and finally, another impulse of about 40 days' range.

In any case, the response of the system is very complex, which is to be expected from the superimposed influence of the recharge induced by the Isbert reservoir, the presence of other springs at a lower level on the northern edge (Balsa Sineu) and the geometric complexity of the environment itself [8]. The complexity of the system's response is also visible in the cross-correlogram (Fig. 5.44), where three peaks are clearly apparent. In summary and in accordance with the cited work, the influence of the Isbert dam, the possible presence of blocks or compartments in the aquifer and the notable thickness of the unsaturated fringe in a large part of the system all contribute to this response.

Photo 5.3   Bolata spring (Tormos) in May 2002 and April 2005 (Photos J. Ballesteros)

5.7   Box 4: Applying the Methods to Bulgarian Springs

5.7.1   Hydrogeological Characteristics of the Areas Studied

The study area comprises the karst massifs of Kotlin, Bistretz-Manish and Nastan-Trigrad [9, 10], which are well-isolated from both a lithological and tectonic point of view (Fig. 5.45).

The Kotel spring is the main discharge point of the **Kotlin massif** [11], which supports dense forest. It has an average altitude of 800 m and a top height of 1044 m. It is part of the Kipilovska syncline, facing north and riding on the Palaeogene flysch, while the Jurassic flysch is superimposed on the massif at its southern edge (Fig. 5.46). The part drained by the spring is of limestone from the Upper Cretaceous, more than 500 m thick and about

Photo 5.4 Bolata spring (Tormos) in May 2002 and April 2005 (Photos J. Ballesteros)

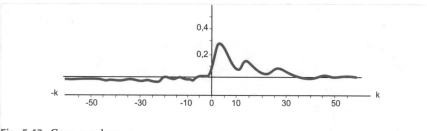

Fig. 5.43 Cross-correlogram

Fig. 5.44 Location of the studied aquifers

30 km² in extent. The limestones are highly karstified, and both exo- and endokarst forms are abundant. The National Institute of Meteorology and Hydrology (INMH) has been measuring spring flow and water temperature since 1962. The spring is used as a water supply. Figure 5.47 shows a hydrograph representing the spring.

One of the main outcrops of the **Bistretz-Manish massif** is near the village of Bistrez. The massif includes the north-eastern part of the Zgori-gradska anticline. Its north-eastern sector is thrust over the Aptian materials of the southern sector of the syncline of Salashka (Fig. 5.48). Both structures are intensely faulted. The average height of the karst massif is 850–900 m, with a maximum height of 1208 m. The massif is made up of limestone and dolomite from the Middle Triassic period. The system is partially isolated

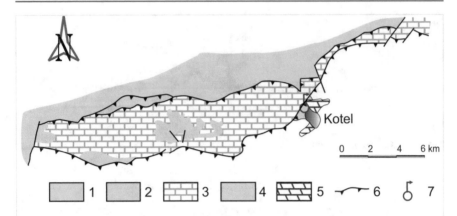

Fig. 5.45 Hydrogeological scheme of the Kotlin karst massif: (1) Alluvial; (2) Limolite, sandstone, marl and flysch (Palaeogene-Eocene); (3) Cretaceous limestone; (4) Sandstone, marl, marl and Jurassic flysch; (5) Limestone and dolomite; (6) Overthrust and reverse faults; (7) Spring

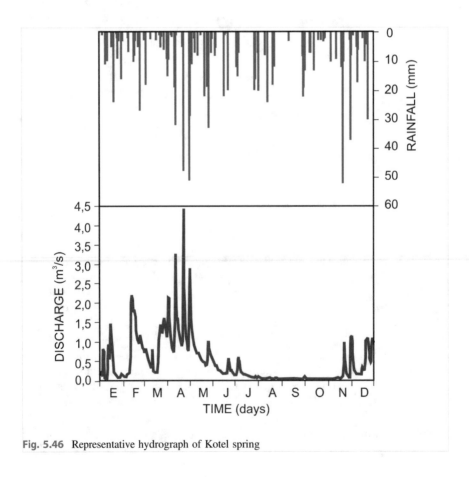

Fig. 5.46 Representative hydrograph of Kotel spring

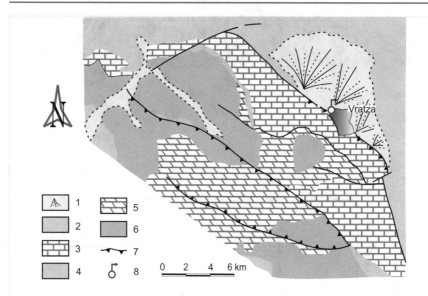

Fig. 5.47 Hydrogeological scheme of the Bistretz-Manish karst massif: (1) Quaternary sediments; (2) Cretaceous sandy marl and marl; (3) Cretaceous and Jurassic limestone; (4) Jurassic shales and limestone (5) Triassic dolomite and limestone; (6) Triassic, Permian and Carboniferous conglomerate, sandstone, argillite and volcanic rocks; (7) Reverse fault; (8) Spring

from the Urgonian material and limestone of the Upper Jurassic, on the one hand, and from the argillite and sandstone of the Lower Jurassic, on the other.

The basin drained by the Bistretz spring is an extensive karst depression of Upper Jurassic limestone covering 24.8 km^2 and more than 500 m thick. The karst development in the massif is remarkable, featuring the Cave of Ledenika [11] the most important one. The flows, temperature and chemistry of the springs are controlled by the Bulgarian Water Administration (NIMH). They are also used as an urban water supply. Figure 5.49 shows a representative hydrograph of the spring.

The spring near the village of Beden is one of the most important ones draining the karst massif of **Nastan-Trigrad**. At almost 300 km^2 in extent, it has an average altitude of 1500 m and is part of the synclinal of the Southern Rhodopes. The Precambrian marble and the outcropping dolomite reach a thickness of about 2000 m (Fig. 5.50). To the east, south and west, the carbonate rocks are in contact with volcanic rocks and crystalline rocks: granite, rhyolite, gneiss and schist of various ages. There are also local conglomerate and Palaeocene sandstones. Table 5.2 summarizes the main characteristics of the three springs studied.

Karstification has intensely affected all the marble. Recharge is achieved by rain infiltration and through the valleys that descend from the marble.

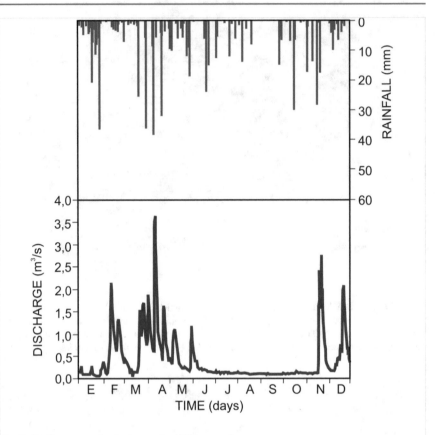

Fig. 5.48 Representative hydrograph of Bistrez spring

These occupy about 62 km². The flows (Fig. 5.51) water temperature and chemistry of the mainsprings are controlled by the NIMH.

5.7.2 Simple Correlation and Spectral Analysis

Precipitation can be considered as white noise and deduced from the auto-correlogram and spectrum (Fig. 5.52a, b). The spectral functions of Kotel and Beden are similar in both their shape and the position of their peaks, while Bistretz is different, perhaps due to the dissimilar climatic conditions. The influence of the Mediterranean and the Black Sea is notable at both Beden and Kotel.

The memory effect estimated from the correlogram for the spring of Kotel varies between 60 and 80 days (Fig. 5.53). The decrease in function is not homogeneous, and it has two components. The first has a duration of 10 days

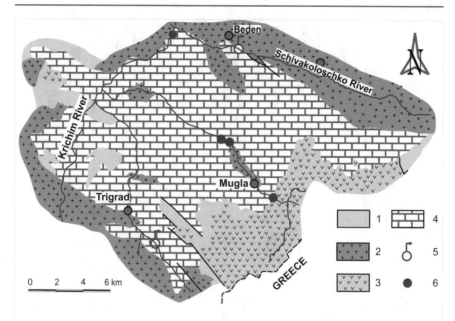

Fig. 5.49 Hydrogeological scheme of the Nastan-Trigrad karst massif: (1) Palaeogene conglomerate and sandstone; (2) Ignimbrite, granite and gneiss; (3) Rhyolite; (4) Precambrian marble; (5) Spring; (6) Sinkhole

and decreases faster than the second, where two other inflections can be detected, in turn, at 24 and 48 days. Everything points to this being due to snowmelt influencing the recharge of the system.

As in Kotel, the Bistretz discharge correlogram shows two components Fig. 5.53a). The first decreases rapidly after 10 days but the second decreases more slowly, reaching zero after 40 or 60 days. This component is somewhat different from Kotel, because of the dissimilar climate.

In addition to the 365-day peak, other peaks are evident (Fig. 5.53b) at 51, 24, 16 and 12 days. These may have various origins, relating to precipitation and the delayed effect of snowmelt. The regulation of the system is low, and fluctuations in the input signal are detected in less than five days.

The spectrum is similar to that of Kotel (Fig. 5.53), but covers a greater range of variance in the high periods and lower in the medium and low periods. Spectrum values for periods of less than five days are practically zero. The peaks in the low frequencies correspond to periods of 365, 51, 20, 16, 12 and 10 days, and their origins are similar to those of Kotel.

The correlogram of Beden's discharge indicates that it has the spring with the greatest memory effect: 70–90 days. Its profile is homogeneous (Fig. 5.53) and without any significant changes. There is only one component. The spectrum shows a remarkable signal transformation capability,

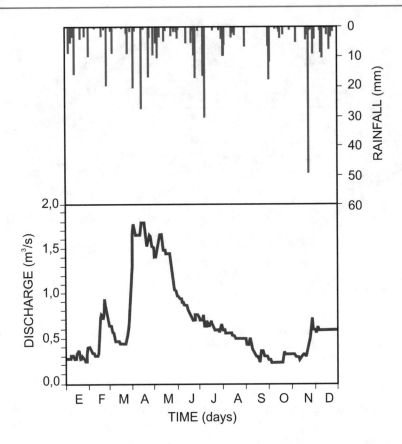

Fig. 5.50 Representative hydrograph of Beden spring

Table 5.2 Basic characteristics of the springs investigated

Spring		Kotel	Bistretz	Beden
Altitude (*m*)		504	302	785
Years		1983–88	1983–89	1985–89
Series length (d)		2192	2557	1826
Discharge l/s	Maximum	8700	4142	1996
	Medium	425	354	628
	Minimum	52	23	240
Rainfall gauge		Kotel	Vratza Varshec	Devin
Average precipitation (mm)		883	716–833	561
Average temperature (°C)		9.1	11.3–1.8	9
Snow (*d*)	Maximum	26	43–84	106
	Medium	6	7–12	13
	Minimum	1	1–1	1

compared to the other two. In the low frequencies, it is visible only in the low
frequencies (long periods; Fig. 5.53).

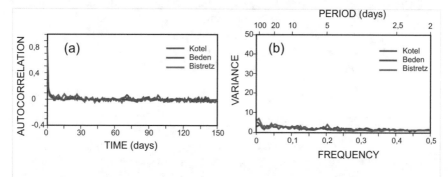

Fig. 5.51 a Autocorrelograms; and **b** spectra of rainfall near karst springs

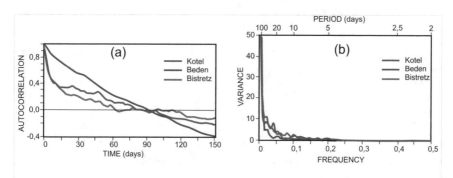

Fig. 5.52 Kotel, Beden and Bistretz spring discharge: **a** correlograms; **b** spectra

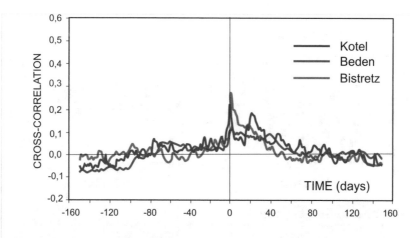

Fig. 5.53 Cross-correlograms, Kotel

5.7.3   Cross-Spectral Analysis

The cross-correlogram of Kotel's daily discharge and rainfall (Fig. 5.54) shows two characteristics of the system: a quick response, with a delay of three days; and a maximum peak at 20 days' and 14–20 days' duration. This is the effect of thawing of the snow cover. The average duration of this snow cover is six days. Consequently, the delays are not due to the system itself but to the effect of thawing, passing into and spreading through the soil, vegetation and the unsaturated belt [10].

The cross-amplitude function is analogous to the gain function (Fig. 5.55). Their values are slightly higher at low frequencies than at medium and high frequencies. However, these frequencies do not have significant values. This is due to the component of the transform function generating a fast response output for the input function. One of the peaks, corresponding to 20–23 days, is well-marked in both functions and can be considered as linked to the snow cover. The delay, estimated using the phase function (Fig. 5.55) in the low frequencies, is of the order of 19 days, which corresponds to the average and maximum   duration   of   snow   cover.   The   estimated   delay   in   the

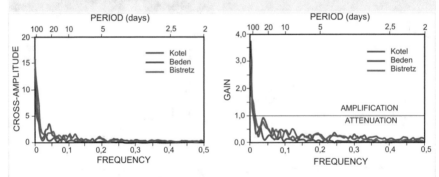

Fig. 5.54 Cross-amplitude and gain functions

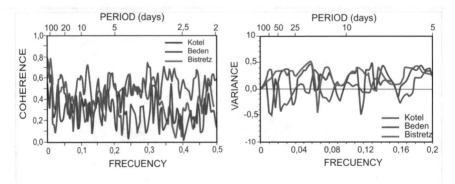

Fig. 5.55 Coherence and phase functions

mid-frequencies in the phase is 2–15 days, as it is not clearly identified. This delay must correspond to rapid circulation in the karst system.

The cross-amplitude function is analogous to the gain function (Fig. 5.55). Both have high values at low frequencies but not at mid- and high frequencies. The peak corresponding to the period 20–23 days is very well-defined in both functions and is a reflection of the retarding effect of snowmelt. The coherence function (Fig. 5.55) is very irregular, with values around 0.5, equivalent to the lack of coherence due to its weak regulatory power.

The Bistretz cross-correlogram (Fig. 5.54) indicates a system memory of between 32 and 36 days. The shape of the curve suggests two types of transformation; the first is a first rapid response, with a day's delay, while the second is relatively low and gradual. Snowmelt is also identified, although not as markedly as in the previous case. The explanation is the duration of the snow cover, for two main reasons. The first is the frequency distribution of its duration. When this frequency has a maximum of several days of duration, a well-marked maximum is also identified in the cross-correlogram; the snow cover transforms the function related to precipitation, as was the case in Kotel. If, on the other hand, the frequency distribution is uniform, the curve of the cross-correlogram shows a continuous descent, as in the Beden spring. Another possible explanation is that the system has a great capacity for regulation.

The cross-amplitude and gain functions (Fig. 5.55) are similar to those of Kotel. The signal variance indicates two types of circulation, one fast and one slow. However, the cross-amplitude and gain functions are not zero in the mid- and high frequencies. The peaks of snowmelt are not well-identified, apart from in the gain function, which is visible in 27 days, very possibly linked to the effect of snowmelt. The coherence (Fig. 5.55) is similar to that of Kotel, although in this case it varies around low values (0.3–0.4). This confirms the hypothesis regarding its high regulatory power. The phase function shows a 19-day delay at low frequencies.

Beden's cross-correlogram is also smooth (Fig. 5.54), more like an intergranular porosity medium than a karst medium. Its regulatory power is high with a memory effect of about 50–60 days. The cross-amplitude and gain functions (Fig. 5.55) show a strong reduction of the input signal at the mid- and high frequencies, to the detriment of the low frequencies, which is typical of well-regulated systems. The estimated delay at low frequencies is 13 days, according to the phase function. Significant phase distortion is observed in the mid- and high frequencies due to the reduced variance of the input function.

5.8 Further Reading

Box, G.E.P., Jenkins, G.M. 1976. *Time Series Analysis: Forecasting and control*. San Francisco: Holden–Day, 575 p.

Brillinger, D.R. 1975. *Time Series Data Analysis and Theory*. International series in decision processing. New York: H.R.W., 500 p.

Chang, Y. et al. 2019. Modelling spring discharge and solute transport in conduits by coupling CFPv2 to an epikarst reservoir for a karst aquifer. *Journal of Hydrology*, 569: 587–599.

Fiorillo, F., Doglioni, A., 2010. The relation between karst spring discharge and rainfall by cross-correlation analysis (Campania, southern Italy). *Hydrogeology Journal*, 18: 1881–1895.

Mangin, A. 1975. Contribution à l'étude hydrodynamique des aquifères karstiques. Doctoral thesis, Dijon. *Ann. Spéléol.*, 29 (3): 283-332; 29 (4): 495–601; 30: 121–124.

Mayaud, C., Wagner, T., Benischke, R., Birk, S. 2014. Single event time series analysis in a binary karst catchment evaluated using a groundwater model (Lurbach system, Austria). *Journal of Hydrology*, 511: 628–639.

Padilla, A., Pulido-Bosch, A. 1995. Study of hydrographs of karstic aquifers by means of correlation and cross-spectral analysis. *J. Hydrol.*, 168: 73–89.

Padilla, A., Dimitrov, D., Pulido-Bosch, A., Machkova, M. 1994. On the application of spectral analysis for investigation the karst spring outflow characteristics. *Bulgarian Journal of Meteorology and Hydrology*, 5 (1–2): 6–19.

Pardo-Igúzqiza, E. et al. 2019. A review of fractals in karst. *Int. J. Spel*, 48–1: 11–20.

Squarzonia, M.F.G., De Waele, J., Vigna, A.F.B. et al. 2018. Differentiated spring behavior under changing hydrological conditions in an alpine karst aquifer. *Journal of Hydrology*, 556: 572–584.

5.9 Short Questions

1. How many components can be identified in a time series?
2. What advantages do you think discharges and rainfall time series analysis has over base flow and recession analysis?
3. How many years of recording are required to interpret system behaviour correctly?
4. What does the correlogram of a time series consist of?
5. What is the density of variance spectrum?
6. What is the difference between a simple correlogram and cross-correlogram?
7. What is the cross-correlogram of a representative hydrological series assimilated by?
8. What is meant by the regulation time of a karst spring?
9. What would a very sharp, simple correlogram indicate, compared to a soft or stretched one?
10. To what extent does a simple correlogram reflect the organization of karst flow? Give reasons for your answer.
11. Does it make sense to draw conclusions about the organization of the flow in a system from the undulations of a simple correlogram? Give reasons for your answer.
12. When working with daily data, why is rain usually considered as white noise?

13. Can monthly rainfall data be considered a purely random process? Give reasons for your answer.
14. Why does the information provided by a simple correlogram depend on the observation window?
15. How are acute and short responses interpreted in a simple correlogram? What about soft and long-distance correlograms?

5.10 Personal Work

1. Simple spectral analysis and its application to the study of karst springs.
2. Cross-analysis in the time domain, as applied to the study of karst springs.
3. Analysis in the frequency domain, as applied to the study of karst springs.
4. Application of simple and cross-analysis to aquifers of great geometric complexity (Fontestorbes and Vaucluse springs).

References

1. Brillinger, D. R. (1975). *Time series data analysis and theory. International series in decision processes* (500 p). New York: H.R.W.
2. Box, G. E. P., & Jenkins, G. M. (1976). *Time series analysis: Forecasting and control* (575 p). San Francisco: Holdenday.
3. Max, J. (1980). *Méthodes et techniques du traitement du signal et application aux mesures physiques* (379 p). Paris: Masson.
4. Mangin, A. (1982). L'approche systémique du karst, conséquences conceptuelles et méthodologiques. *Reunión Monográfica Karst Larra*, 141–157.
5. Padilla, A., & Pulido-Bosch, A. (1992). Consideraciones sobre la aplicación de los análisis de correlación y espectral al estudio de los acuíferos kársticos. *Taller Internacional sobre cuencas experimentales en el karst* (pp. 149–160). Playa Girón, Cuba.
6. Padilla, A., & Pulido-Bosch, A. (1995). Study of hydrographs of karstic aquifers by means of correlation and cross-spectral analysis. *Journal of Hydrology, 168*: 73–89.
7. Benavente, J., Mangin, A., & Pulido-Bosch, A. (1985). Application of correlation and spectral procedures to the study of discharge in a karstic system (Eastern Spain). Cong. Intern. Hydrogeol. Karst. Ankara. In *Karst water resources* (Vol. 161, pp. 67–75). IAHS.
8. Pulido-Bosch, A., & Benavente, J. (1987). Contribución de la deconvolución al estudio de la descarga de la Unidad Alfaro–Mediodía–Segaria (Alicante). In *IV Simposio de Hidrogeología y Recursos Hidráulicos*, (Vol. XI, pp. 411–420). Mallorca.
9. Padilla, A., Dimitrov, D., Pulido-Bosch, A., & Machkova, M. (1994). On the application of spectral analysis for investigation the karst spring outflow characteristics. *Bulgarian Journal of Meteorology and Hydrology, 5*(1–2), 6–19.
10. Pulido-Bosch, A., Padilla, A., Dimitrov, D. & Machkova, M. (1995). The discharge variability of some karst springs in Bulgaria studied by time series analysis. *Hydrological Sciences Journal, 40*, 517–532.
11. Antonov, H., & Danchev, D. (1980). Ground waters in Bulgaria. In *Technika* (p. 359). Sofia (in Bulgarian).

Mathematical Models

6

6.1 Glossary

Black-box model: The black box is the aquifer; precipitation or ***useful rain*** is the *input function* and the *output function* is a spring's flow.

Convolution and deconvolution: When a system is linear (proportionality and additivity between inputs and outputs) and invariant (the properties of the black box do not depend on time and are independent of input and output), the black box is represented by the Duhamel integral or convolution integral.

The ***impulse response or convolution core*** is calculated from the data of inputs and outputs, using ***deconvolution*** in which n is the system ***memory***.

The ***norm*** or ***residue*** (quadratic difference between the measured values and those obtained by convolution) is the criterion that allows us to know the goodness of fit of the generated series.

Precipitation-discharge models have as their input the rainfall over the basin, and as output the flow measured in the spring.

Univariate models are based on analysis of the stochastic structure of time series. Among them is the integrated *auto-regressive moving averages* (ARIMA).

ARMAX model (***p***, ***q***, ***r***), *auto-regressive moving average with exogenous variables*, part of y_t, the series to be simulated, dependent on the xt or known exciter series.

'Specific' models are those based on the conceptual karst aquifer model of *blocks and conduits*. The ***adjustment criteria*** are usually the measured and simulated flows of a spring and the reproduction of piezometric evolutions.

© Springer Nature Switzerland AG 2021
A. Pulido-Bosch, *Principles of Karst Hydrogeology*, Springer Textbooks in Earth Sciences, Geography and Environment,
https://doi.org/10.1007/978-3-030-55370-8_6

6.2 General Aspects

Within the Spanish karst aquifers, we find a variety of types as the Calar del Mundo, with its spectacular waterfall on an underground river, and the Sierra de Crevillente, where almost all boreholes have high yields, or the case of the *ufanas* of the Sierra Norte in Mallorca. These change in less than 90 h from being dry to issuing 19 m^3/s, then dry up again. Aquifers cannot be simulated for springs such as these *ufanas* or those of Aliou in the French Pyrenees (peaks of 20 m^3/s and a base flow of 10 l/s), or Deifontes, where the fluctuation in the flow is less than 30% of the average discharge.

Hydrographs of karst springs have always been the object of special attention in the study of this type of aquifer, although in general, only a small part of the hydrograph (depletion, recession or global) is used in calculations and, in these cases, a considerable amount of information remains unprocessed. The procedures usually used for analysis of such time series can work with the whole hydrograph to obtain all the information hidden in them. In addition, they allow the system to be characterized and can serve as a starting element in choosing the method to simulate the system. These procedures are correlation and spectral analyses, both simple and crossed.

The correlogram of the discharge of the spring, in comparison with that for rainfall in the same area, provides information on the regulatory power of the aquifer system and the importance of its reserves. Its regulation time is also quantifiable. According to the form of the correlogram function, the inertia of the system can be specified: the higher the degree of correlation between the various flows of the series, the more similar is the karst aquifer system to an aquifer of intergranular porosity.

The variance density spectrum represents the distribution of the variances of the time series for the different frequencies. Comparison of the spectrum of the input function (rainfall) with that of the output function (discharges) makes it possible to recognize which variations the system filters and which it transmits. In our latitudes, in general, only low frequencies remain unfiltered. This means that the long-period variations in precipitation are reproduced by the flows (droughts and periods of recharge), while an isolated rain shower is not reflected in the spectrum of discharge due to the system's inertia. The cross-analysis characterizes the relationship between a cause (rainfall) and an effect (discharge at a spring). Of all the possibilities supported by cross-analysis, cross-correlogram must be paramount, as it provides an accurate representation of the system's impulse response.

6.3 Black-Box Models

This type of model considers the input and output functions to be separated by a 'black box', which is where the transformation of the input function into the output function takes place. The black box is the aquifer; the input function is rainfall and the output function is a spring's flow.

There are many calculation models or procedures under this denomination. Deconvolution, deposit models and ARMA, ARIMA and ARIMAX are some of the most relevant, as described below.

6.3.1 Convolution and Deconvolution

When a system is linear (has a proportionality and additivity between inputs and outputs) and invariant (the properties of the black box do not depend on time and are independent of input and output), the black box is represented by the Duhamel integral, or convolution integral:

$$S(t) = \int_0^\infty R(\tau)E(t - \tau)\mathrm{d}\tau 1 \qquad (6.1)$$

where $S(t)$ is the system output, $E(t)$ the input, $R(\tau)$ the convolution core and t the time. The impulse response, equivalent to the unit hydrograph, represents the output corresponding to a unit input. The impulse response is calculated from a series of input and output data, using what is known as ***deconvolution***. The calculation is generally made on a series of daily rain values and flow data of at least five years.

When the system is neither linear nor invariant, the solution to the problem is a little more complicated, although it has a solution. Something similar happens when there is more than one entry. Multiple linear regression can also be applied to the resolution of black-box models, although the cause–effect relationship has less obvious physical significance.

6.3.2 Tank Models

This type of model, as applied to karst, is derived from earlier applications to reconstruct the flows measured by gauging stations in surface basins. They are usually known by the generic term, rainfall-discharge models, although they have diverse names alluding to the centre where the algorithms were developed (Stanford model), to the author who conceived them (Girard model [1]; Mero model [2]), or are simple acronyms (CREC model [3]; BEMER model [4]; or TRIDEP model [5, 6]).

The principle of these models is the same. The input element is the rainfall over the basin, and the output element is the flow measured at a point in a river or at a spring. Rainfall is distributed in series and/or parallel boxes that aim to simulate evapotranspiration, run-off, the buffering effect of the unsaturated zone, the saturated zone and the turbulent and/or laminar flow within it. Let us briefly discuss the Mero, CREC, BEMER and TRIDEP models.

6.3.2.1   Mero Model

This Mero model has at least eight versions, each an improvement on the previous, and was developed by the Ministry of Agriculture of Israel. It consists of four tanks: U, L, R_3 and R_4. The first represents the soil and is supplied by evapotranspiration, passing the excess to surface run-off and/or to the deposit L. This new tank represents the aeration fringe, and is in turn divided into two subdeposits, one representing capillary storage (it feeds actual evapotranspiration when the U tank is exhausted), and the other feeding the saturated fringe, depending on the absorption capacity of the soil. When the two subdeposits are full, the excess feeds the subsurface flow run-off. The water from L passes to the two remaining tanks, R_3 and R_4, which represent the saturated strip and have a different emptying regime. A total of ten parameters need to be determined for model fitting, performed by fitting to known data on daily rainfall and discharge series.

6.3.2.2   CREC Model

The CREC model simulates several karst springs, and it obtains satisfactory results. The general operation of the model is summarized in Fig. 6.1. As shown, it consists

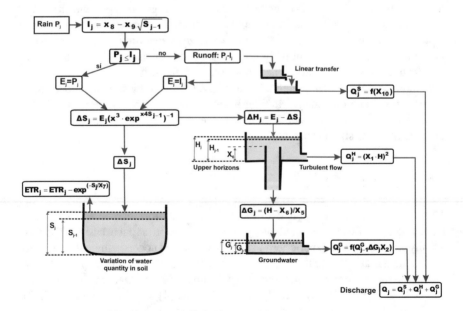

Fig. 6.1 CREC model [3]. Pj: daily rainfall (input); Ij: potential infiltration; Ex: infiltrated rain; ΔSj: tank feed S; X_3, X_4, X_7: tank parameters S; Qj: surface run-off; ΔHj: tank feed H; X_6: potential storage threshold; QjH: output from tank H, function of parameter X_1 and level of tank Hj; ΔGj: supply to tank G from H: QjG: discharge flow from tank G; Qj: global flow from the basin (output of the model, to be compared with the measured flow)

of three tanks S, H and G, representing the soil and the saturated zone, with a nonlinear and a linear emptying regime, respectively. There are seven parameters that regulate the model, obtained by iterative calculation.

6.3.2.3 BEMER Model

Unlike the other two models, this was designed to simulate water transfer in a karst aquifer. There are five simulated boxes, the first representing the soil, and feeds surface run-off and/or evapotranspiration (Thornthwaite method) and/or infiltration. The infiltrating water passes to three deposits placed in the unsaturated strip, the purpose of which is to create a gap between entry and exit that reflects the various infiltration capacities of the land. The water from these three tanks is transferred to the last tank, located in the saturated zone. Here, depending on the sheet of water existing, a coefficient of depletion of a greater or lesser value is used to empty it. From this last tank, water can also be extracted by pumping.

The input data for the model are the monthly mean temperatures (for calculation of potential evapotranspiration), rainfall and daily flows (from springs and/or pumping). In the model adjustment, 22 parameters are involved, optimized by successive non-automated tests. The results obtained in the reconstruction of karst spring flow series are satisfactory, although the large number of parameters involved in the adjustment of the model makes it unwieldy.

6.3.2.4 TRIDEP Model

Theoretical aspects
The TRIDEP rainfall–discharge model comprises three zones, each characterized by a transfer function (Fig. 6.2). The first symbolizes the surface zone, where the processes of evapotranspiration and run-off take place, which is then subtracted from the precipitation to obtain the effective infiltration. The second is the unsaturated zone with a delay effect on infiltration, and the last is the saturated zone whose output function characterizes the decreasing curve of the hydrograph. The following is a brief description of the functions that define the hydraulic behaviour of each of these zones.

In Fig. 6.2, VI_{i-t} = volume of effective infiltration over time $i - t$; VA_i = volume added to the saturated zone at instant i; λ_t = parameters of the convolution kernel. In a discrete form and assuming that the convolution parameters vanish after interval n, the integral can be expressed as: (Fig. 6.3).

$$VA_i = \sum_{j=0}^{n} VI_{i-j}\lambda_j \tag{6.2}$$

Parameters can be estimated by the Fourier transform method or by solving the resulting system of equations by minimizing the quadratic error between the calculated and theoretical volumes or any other. Q_i can be obtained by the following function:

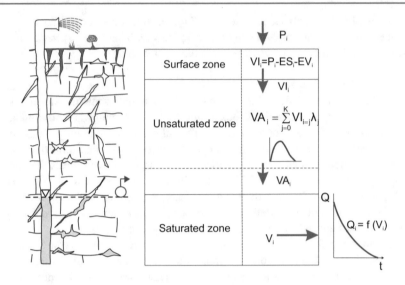

Fig. 6.2 Diagram showing the operation of the three zones used in the model

Fig. 6.3 Scheme showing
the procedure for calculating
the volume contributed to the
saturated zone

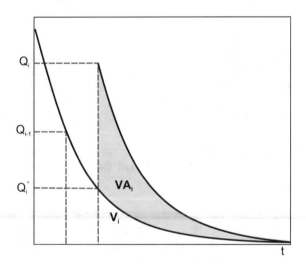

$$Q_i = BQ_{i-1} + \alpha \sum_{i=0}^{n} VI_{i-j} \lambda_j \tag{6.3}$$

which gives the discharge in time i as a function of the previous discharge, of the depletion coefficient and of the effective infiltration in the previous n instants. The presence of pumping activity in the aquifer system can be contemplated by the above equation, so that the new expression is:

$$Q_i = BQ_{i-1} + \text{VB}_i + \sum_{j=0}^{k} \text{VI}_{i-j}\,\lambda_i \tag{6.4}$$

where VB_i = pumped volume in the i interval.

If effective infiltration was a linear phenomenon, with respect to rainfall, it would suffice to replace it in the previous equation with the latter. Since the volume of water infiltrating into the aquifer is not proportional to rainfall but also depends on seasonal phenomena (temperature and luminosity), intensity and soil moisture conditions, it is necessary to estimate it by empirical methods (Thornthwaite, Kessler, Coutagne or others). These methods are what we consider as forming the transfer function of the surface zone. Some examples are described in Box 1.

6.4 The AR, MA, ARMA and ARIMA Processes

6.4.1 ARIMA Models

Univariate models are based on the analysis of the stochastic structure of a time series to forecast complete data in the short term, or to generate a synthetic series that preserves the main statistics, especially the temporal correlation (autocorrelation function) of the underlying process in the original series. Box and Jenkins in Chap. 5 [2] studied in detail models that, in general, are known as the autoregressive integrated moving average (ARIMA).

The purpose of ARIMA models is to obtain a linear filter that converts a sequence of random variables or white noise, a_t, uncorrelated, into a sequence of variables, z_t, correlated. If we consider the stationary time series z_t centred and normally distributed, which has an autoregressive correlation (that is, a time-dependent structure following a Markovian process), the autoregressive model of order p, AR (p), which represents the variable, can be written:

$$zt = \Phi_1 z_{t-1} + \Phi_2 z_{t-2} + \cdots + \Phi_p z_{t-p} + a_t \quad \text{or} \quad z_t = \sum_{j=1}^{p} \phi_j\, z_{t-j} + a_t \tag{6.5}$$

where Φ_j constitutes the parameters of the autoregressive model that defines the linear filter and a_t a random variable that follows a certain distribution, generally normal. The parameters necessary to define an AR model (p) are the mean, variance and p autoregressive parameters; a total of $p + 2$.

If the time series zt depends only on a number, q, finite of random variables, at, prior to the time instant t, the result is a process of moving averages of order q, MA (q). This is expressed as:

$$z_t = a_t - \theta_1 a_{t-1} - \cdots - \theta_{qat-q} \quad or \quad z_t = a_t - \sum_{j=1}^{q} \theta_j a_{t-j} \qquad (6.6)$$

where θ_j makes up the set of parameters of the moving average model. The total number of parameters required to adjust an MA(q) model is $q + 2$.

The combination of an autoregressive model of order p and a model of moving averages of order q will result in the autoregressive model of moving averages of order p, q, ARMA(p, q), which will be defined with a total of $2 + p+q$ parameters. Its general expression is:

$$zt = \Phi_1 z_{t-1} + \Phi_2 z_{t-2} + \cdots + \Phi_p z_{t-p} + a_t$$
$$- \theta_1 a_{t-1} - \theta_2 a_{t-2} - \cdots - \theta_q a_{t-q} \quad or$$
$$z_t = \sum_{j=1}^{p} \phi_j z_{t-p} + - \sum_{j=1}^{q} \theta_j z_{t-q} \qquad (6.7)$$

When z_t is not stationary, essentially in the average, due to a manifest secular trend, the variable can be transformed into a new one by means of a differentiation operation of order d, in order to suppress the trend and make it stationary. If $d = 1$, then $u_t = z_t - z_{t-1}$. In general, a differentiation of order d in z_t will give rise to a new variable u_t that will be expressed by:

$$u_t = z_t - z_{t-1} - z_{t-2} - \cdots - z_{t-d} \qquad (6.8)$$

The process of differentiation in an ARMA(p, q) model results in the more general ARIMA (p, d, q) model. The seasonal periodicity, so ubiquitous in hydrological series, can be suppressed by introducing a differentiation with a delay ω, equal to the length of the periodicity. For example, if the length of the sampling interval of the u_t series is monthly and shows an annual periodicity, it can be transformed into a new series, ω_t, without periodicity, with a differentiation of order $\omega = 12$, in the following way:

$$\omega_t = u_t - u_{t-12.} \qquad (6.9)$$

Then, we have the integrated autoregressive model of seasonal moving averages:

$$ARIMA(p, d, q)_\omega. \qquad (6.10)$$

When the series of residues remaining when fitting an ARIMA(p, d, q)$_\omega$ model is not random but have a certain self-regressive structure dependent on each other, a new linear filter can be applied that takes the form of an ARIMA(P, D, Q)$_\omega$ model. This is the general model of Box and Jenkins in Chap. 5 [2] Auto-Regressive Integrated of Moving Averages multiplicative, ARIMA(p, d, q) $\times$ (P, D, Q)$_\omega$.

The properties of the ARIMA models and their demonstration, as well as the spatial limitation of their parameters, can be consulted in a bibliography on the subject. With a view to its practical application, it should be pointed out that two of the main conditions to be met by an ARIMA model are that the series to be modelled must be stationary and normally distributed, that the resulting residues must be independent of each other.

The functions used to describe and characterize an ARIMA model are those mentioned in the section on correlation analysis and simple spectral analysis, of which the autocorrelation function (ACF) is particularly significant. However, it is the partial autocorrelation function (FAP) that best identifies this type of model.

The FAP coefficients represent a measure of the linear association between the FAC coefficients [7]. The graphical representation of the function reflects the dependency structure of the autoregressive coefficients. For example, in the simplest case where only the first parameter, c_1, of the FAP is significantly different from zero, the r_k coefficients of the FAC would be determined by $r_k = c_1 r_{k-1}$.

This process represents an AR(1) model with $\Phi_1 = c_1$, and is hydrologically equivalent to the linear discharge of a spring, so that $Q_t = cQ_{t-1}$, where c can represent, for example, an exponential function of the $e^{-\alpha \Delta_t}$ type. The FAP coefficients are obtained by solving the Yule-Walker system of equations: $[R_k] C_k = R_k$

$[R_k]$ represents the symmetrical square matrix $(k \times k)$ of the FAC coefficients, C_k the vector dimension column k of the FAP coefficients and R_k the vector column of the FAC coefficients. Three classic methods are used to estimate the parameters of an ARIMA model: the method of moments, the method of maximum probability, and the method of least squares [8, 7], in Chap. 5 [2]. The least squares method consists of minimizing the quadratic error function, FE, which results from the difference between the experimental values, z_t, and the estimates, $z_t^\wedge$; the following is written:

$$\text{FE} = \sum_{t=1}^{N} (z_t - \hat{z})^2 = \sum_{t=1}^{N} a_t^2 \tag{6.11}$$

N is the available number of experimental values and a_t the residues remaining when applying the model. $\hat{z}_t$ is a function of the estimated parameters c_k and the previous values $\hat{z}_t = f(z_{t-1}, z_{t-2}, \ldots, c_1, c_2, \ldots, c_k)$.

In order for FE to be minimum, its first partial derivative with respect to each of the parameters must be equal to zero. Marquardt's iterative algorithm [9] can be used to solve the above equations.

Similar to the development made by Salas et al. [7] to physically justify an ARMA model applied to the flow of a river linked to an aquifer, an attempt could be made on the discharge of a spring draining a karst system. This could be the karst aquifer outlined in Fig. 6.4, where the variables represent discrete values taken in an interval. Let us also consider that in this system, the functions that link the transfer of volumes are linear; that is, invariant in time and independent of the state of the system. Precipitation at a given instant, P_t, will be distributed between

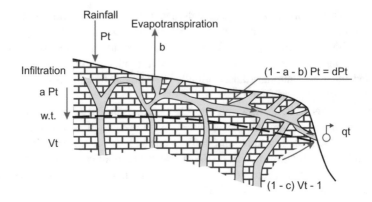

Fig. 6.4 Conceptual representation of karst spring discharge processes. Adapted from [7]

infiltration into the saturated zone, aP_t; evapotranspiration, bP_t; and infiltration circulating rapidly through the large ducts, if the system is isolated, will be equal to $(1 - a - b)P_t = dP_t$. Parameters a, b and d represent the various fractions into which the precipitation is divided.

6.4.2 Univariate ARMA Models

Let y_t be the series to be simulated, which depends on the known x_t or exciter series. The model called ARMAX (p, q, r), as an abbreviation of *auto-regressive moving average with exogenous variables*, is expressed by:

$$y_t = \sum_{i=1}^{p} \phi_i\, y_{t-i} + \sum_{j=0}^{p} \omega_j \quad x_{t-j} \sum_{k=0}^{r} \theta_k \quad a_{t-k} + a_t \tag{6.12}$$

The starting hypothesis is that there is a linear relationship between the input and output series. In general, y_t and x_t are series that result from the typification of the series of discharges and rainfall values, respectively, therefore of mean zero and standard deviation unit. Φ_i, ω_j and θ_k are the parameters that characterize the system. When considering the estimated mean and standard deviation, the total number of parameters is $p + q + r + 2$. The variable a_t is white noise with a mean of zero and standard deviation unit, thanks to parameter Φ_0. It is a random variable not combined with itself and independent of both y_t and x_t.

One of the objectives of this type of model is the short-term prediction of the output variable. The improvement in prediction over ARIMA models is that, in the case of univariate ARMAX, there is knowledge of the stochastic structure of the exciter variable as well as the output variable and its interrelation; and in the case of multivariate variables, of the set of input series. In recent years, with the development of Kalman filters and Bayesian techniques applicable to ARMAX models, there has been a notable improvement in real-time prediction. The application of ARMAX to the modelling of karst systems will simulate the series of spring discharges, based on the stochastic structures of the series for calibrating the model, with rainfall values as the known input data.

The function impulse response, which relates input to output, can be obtained by successively developing equations for the various values of y_{t-1}, y_{t-2}, etc. Thus, for example, the ARMAX model $(2, 1, 0)$ can be written as:

$$y_t = \Phi_1 y_{t-1} + \Phi_2 y_{t-2} + \omega_0 x_{t-1} + \omega_1 x_{t-1} + a_t \tag{6.13}$$

This equation is satisfied for any value of t, therefore:

$$y_{t-1} = \Phi_1 y_{t-2} + \Phi_2 y_{t-3} + \omega_0 x_{t-1} + \omega_1 x_{t-2} + a_{t-2}$$

to be replaced by (6.13):

$$\begin{aligned} y_{t-1} = {}&(\Phi_2 + \Phi_1)^2 y_{t-2} + \Phi_1 \Phi_2 y_{t-3} \\ &+ \omega_0 x_t + (\omega_1 + \Phi_1 \omega_0) x_{t-1} + \Phi_1 \omega_1 x_{t-2} \\ &+ a_t + \Phi_1 \, at_{-1} \quad t_{-1} 1 \, t_{-2} \end{aligned}$$

By successively replacing the values of y_{t-2}, y_{t-3}, and so on, one more term is added in x_{t-i} and one less in y_{t-1}. The coefficients multiplying a y_{t-j}, for $j > i$, are increasing powers of Φ_i, which will tend to zero, being a stable system; the independent and random terms appearing from a_{t-i} can be grouped into one generic a_t, independent and random at the same time. Finally, if the coefficients that multiply x_{t-i} are designated by v_i, the model will be expressed as: $v_0 x_t + v_1 x_{t-1} + v_2 x_{t-2} + \cdots + a_t$ or

$$y_t = \sum_{i=0}^{\infty} v_i x_{t-1} + a_t \tag{6.14}$$

V_i coefficients are referred to as the function impulse response and represent the unit hydrograph of the system with Dirac function characteristics. Expression (6.14) is equivalent to a discretized convolution which, by due of its properties, can be written:

$$y_t = \int_{i=-\infty}^{t} v_{t-i} x_i \, di \tag{6.15}$$

where V_t invariant in time, represents the nucleus of convolution or kernel. If b indicates the delay in system response, the Yule-Walker equations relate the coefficients of the function impulse response to those of the ARMAX model, so that

$$v_i = 0 \qquad\qquad\qquad\qquad\qquad\qquad i < b$$
$$v_i = \Phi_i v_{i-1} + \Phi_2 v_{i-2} + \cdots + \Phi_p v_{i-p} + \omega_{i-b} \quad i = b, b+1, \ldots, b+q \qquad (6.16)$$
$$v_i = \Phi_i v_{i-1} + \Phi_2 v_{i-2} + \cdots + \Phi i v_{i-p} \qquad\quad i > b+q \quad i > b+q$$

These equations can be used both in one direction and the other, either to obtain the impulse response function from the coefficients Φ_i and ω_i, or vice versa. It is evident that the coefficients θ_k do not intervene in this formulation, since they represent only the stochastic structure of the term white noise a_t. If we multiply both members by x_{t-k} the model raised in expression (6.14), then $y_t x_{t-k} = v_0 x_t x_{t-k} + v_1 x_t$ $_{-1}x_{tk} + \cdots + a_t x_{t-k}$ by taking expected values in the terms of the expression, and considering that x_t and y_t are typified and that a_t is independent of x_t, the relationship of impulse response function with the autocorrelation of x_t, r_x, and the cross-correlation of y_t with x_t, r_{yx}, will have to be given by $r_{yx}(k) = v_0 r_x(k) + v_1 r_x$ $_{(k-1)} + \cdots$.

Generically, and taking into account that v_k is considered negligible from $k + 1$, it is possible to write in matrix form:

$r_{yx}(0)$		$r_x(0)$	$r_x(1)$	$\ldots$	$r_x(k)$		v_0
$r_{xx}(1)$		$r_x(1)$	$r_x(0)$	$\ldots$	$r_x(k-1)$		v_1
.	$=$	.	.	.	.		
.		.	.	.	.		
$r_{yx}(k)$		$r_x(k)$	$r(k-1)$	$\ldots$	$r_x(0)v_k$		

The coefficients of the function impulse response can be obtained by solving the previously proposed system. If the function x_t is independent of itself; that is, it responds to a random function, the equations are expressed as follows:

$r_{yx}(0)$		v_0
$r_{yx}(1)$		v_1
.	$=I$	
.		
$r_{yx}(k)$		v_k

where I indicates the $k \times k$-dimensional identity matrix. In this expression, the impulse response is equivalent to the cross-correlogram between the output and input functions. In that case, the discretized rain in short intervals of time present an autocorrelogram that could be considered null for $k \neq 0$, therefore it can be considered, in applications of the model ARMAX to the karst systems, that x_t is independent of itself.

There are several methods for obtaining the coefficients in the calibration of ARMAX models; the process followed in subsequent applications practically coincides with that proposed by Box and Jenkins in Chap. 5 [2]. An initial estimation of the autocorrelation parameters is taken from those obtained by applying the univariate models type ARIMA to the series of flows. Assimilating that the cross-correlation function, $r_{yx}(k)$ equals the impulse response function, vk; the parameters of moving averages of rainfall, ω_j, are initially calculated from the Yule-Walker equations, which coincide with the estimation by the method of moments. In order to obtain the definitive set of parameters involved in the model, the method of least squares can be followed. Coefficients are obtained by minimizing the FE function:

$$\min_{\phi_i,\,\omega_j,\,\theta_k} \text{FE} = \min \sum_{t=s+1}^{N} a_t^2 \quad s = p + q + r \tag{6.17}$$

The algorithm used has been that of parameter sensitivity, which consists of the iterative approximation by solving the resulting system of equations by developing, in a Taylor series, the first partial derivative of the function with respect to each of the coefficients and equating to zero:

$$\delta\,\text{FE}/\delta\phi_i = \delta\,\text{FE}/\delta\omega_j = \delta\,\text{FE}/\delta\theta_k = 0 \tag{6.18}$$

Since the estimates by the method of the moments are already close to the optimal ones, the convergence is very fast.

Some examples are developed in Box 1.

6.4.3 Box 1: Application of Black-Box Models

To illustrate the application of deconvolution to karst aquifers, we will use several examples: the springs of the Sierra de Sagra (Alicante); the Fuente Mayor de Simat de Valldigna; and the town spring that drains the Torcal de Antequera. The application of deconvolution to the study of karst aquifers may provide interesting information on the characterization of the system [10, 11]. The data were processed through the GMDUEX program of the Centre d' Informatique Geologique de l'Ecole des Mines de Paris. The procedure is based on the discretization of Duhamel's integral. The impulse response is adjusted in a time interval between zero and n, where n is the system **memory**. For the program requirements, n only takes power values of 2 (2, 4, 8, 16, 32, 64 and so on), which is, to a certain extent, a restriction. Once the convolution nucleus has been identified, we can reproduce a series of flows from the nucleus convolution product entered into the system. The criterion that allows to know the goodness of fit of the generated series is the **norm** or **residue** (quadratic differences between the measured values and those obtained by convolution).

Sierra de Sagra

A total of four simulations were carried out in Chap. 5 [8], details of which are shown in Table 6.1. In the first test, a memory effect of four months was considered, no maximum was imposed on the unitary hydrograph, and the adjustment was carried out on only 19 of the 84 months of the series (Months 5–25). The impulse response obtained shows a pronounced maximum in the first month and decreasing values for the remaining three; the fact that the zero value is not reached in this period shows that the system memory is greater than four months. For the same reason, the reconstruction of the entire series from the product of convolution core entered into the system is not very good, especially for low-water periods. The rule obtained is the highest of all. All this was taken into account in the following test, and a maximum was imposed on the response in the first month, the memory effect was extended to eight months and the whole series was used for the adjustment. A decrease in the value of the standard was the result. The memory effect does not extend to eight months, but is annulled by the sixth month. The measured, convo-luted and input series are shown in 6.4. The setting is acceptable, even though the extreme value is not reproduced properly.

The following two tests were performed, considering useful rain as that entering the system. In the first, the maximum was maintained from the response and the memory effect of eight months. Curiously, in this case, the response is not annulled at the end of that period, and is the reason why the norm does not improve with respect to the last test. In the fourth trial, the memory effect was increased to 16 months. After 11 months, the value of zero was reached, a period that would correspond to the memory effect of the system if considering useful rain as the input. The reconstruction of the series is in this case much more acceptable (Fig. 6.5), although some periods remain less well adjusted.

An important datum to identify by means of this technique is the impulse response of the system. The unit hydrographs obtained are shown in Fig. 6.6. In the first two, the input to the system was considered to be the rainfall measured in Denia, and in the remaining two, the useful rainfall was

Table 6.1 Characteristics and results of the four tests carried out (cs complete series)

No.	Entry n	Adjustment interval	Calculation phases	Memory effect (months)	Norm mm^2
1	Rain	5–24	3	4	22.9
2	Rain	cs	4	8	17.9
3	Useful rain	cs	4	8	19.5
4	Useful rain	cs	5	16	9.8

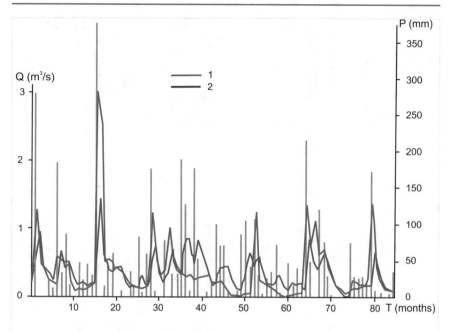

Fig. 6.5 Results obtained in the fourth test: (1) measured values; (2) convoluted values

estimated, discounting the actual evapotranspiration, as obtained by the Thornthwaite method. The differences are remarkable.

From this, we can conclude that the application of the deconvolution to the monthly data of precipitation and flows of the Sagra springs allows us to identify the impulse responses of the system, obtaining various memory effects according to whether gross rain (six months) or useful rain (11 months) is considered as the input to the system. Remember that the deconvolution and subsequent convolution allow us to generate a series, to establish or estimate the real or most probable input, to reconstruct data missing from records and to reproduce a series influenced by human activities —pumping, for example, among possible others. However, the results obtained, relative to memory effect, clearly contradict those obtained from simple and cross-correlation analysis of daily data. This fact has been confirmed by studies carried out on the spring of La Villa (draining Torcal de Antequera) and the Fuente Mayor of Simat de Valldigna (Valencia [12]). The explanation for this observation may be the mathematical method itself, also the physical phenomenon. On the one hand, correlation analysis gives more weight to floods and their relationship to nearby rainfall. On the other hand, in deconvolution, it is the lowlands that largely influence the memory effect, and this is especially evident when it is useful rain that is used as an input function.

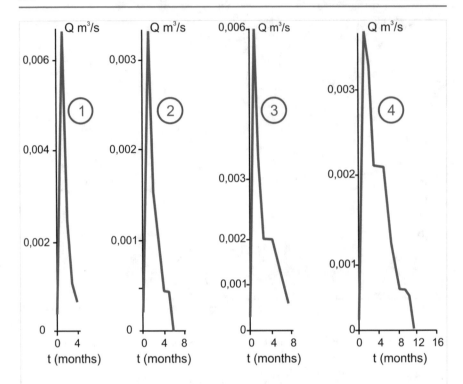

Fig. 6.6 Impulse responses in the four tests

From the physical point of view, it is necessary to qualify that, when dealing with average monthly flows and monthly precipitations, the synchrony of input–output is not always correct. Heavy precipitation at the end of a month will increase the average flow of the subsequent month, normally. Estimating useful rain by Thornthwaite's method monthly also has an effect, given that in months of low water, it is usually zero. This does not necessarily reflect reality, as there may have been intense rain falling over a short interval of time that will have contributed to the system. To reproduce dry months, the impulse response has to be more prolonged.

Finally, as an explanation for the disparity between the data measured and those calculated in the most recent years, it is worth noting the influence of pumping in the vicinity of the springs, as this increased in the latter simulated years.

Fuente Mayor de Simat

The Fuente Mayor in Simat de Valldigna (Valencia) is one of the drainage points of the Sierra Grossa hydrogeological unit in Chap. 2 [15] whose surface area is 370 km^2. Its hydrogeological watershed is not known in detail, but it is estimated to be of the order of 20 km^2. The aquifer materials

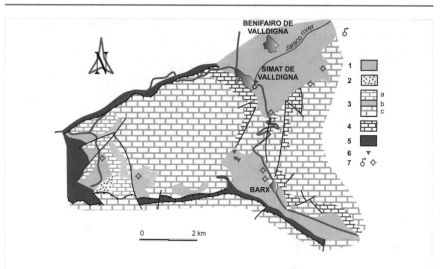

Fig. 6.7 Hydrogeological scheme of Fuente Mayor de Simat de Valldigna and its surroundings: (1) Quaternary detrital deposits; (2) Miocene conglomerate; (3) Creu formation; (4) Jaraco formation; (5) Valencia group; (6) Ponor; (7) Spring and borehole

(Fig. 6.7) correspond to the Creu and Jaraco formations. In this sector, the Creu formation (Cenomanien-Senonien) is represented by a thick dolomitic succession with limestone towards the bottom, an intercalation of about 40 m of sandy marls with irregularly shaped bodies of siliceous gravel, and at the bottom a white limestone. The Jaraco formation (Middle Cretaceous) has integrated limestone and dolomite with marly intercalations. The Keuper's multicoloured clay, piedmont deposits and detrital materials also occur in the Simat valley and in the La Drova-Barx sector. Very close to the source is the polje of Barx, La Drova, with a watershed of 8 km^2. This has an impact on the operation of the spring, a fact known since ancient times ([13, 14, in Chap. 2 [19]). Several sinkholes inside the polje, among which La Doncella stands out, quickly absorb the run-off after heavy rainfall.

The data used correspond to total monthly precipitation and average monthly discharge for a total of 104 months between February 1973 and September 1981; that is, slightly less than nine years. A total of four tests was carried out. Table 6.2 shows the most representative characteristics of each, as well as the average norm of the last calculation phase, referring to the adjustment period, in order to allow the data to be fully comparable. Figure 6.8 also shows the various impulse responses obtained.

The results of the first test are shown in Fig. 6.9. As indicated in Table 6.2, the system memory effect was considered equal to four months and no maximum was imposed. From its form (Fig. 6.9), it can be deduced that this is greater than the period mentioned, since it was not annulled.

Table 6.2 Main characteristics of the various tests carried out; in all cases, the input was raw rainfall, apart from 4, which used useful rain (CFU last calculation phase)

No.	Impulse response (days)	Calibration period (days)	UFC (LCP)	Maximum (days)	Simulated period (days)	Mean CFU (days^2)
1	4	5–24	3	without	1–104	0.127
2	8	9–52	4	1	1–104	0.033
3	8	1–104	4	1	1–104	0.013
4*	16	1–104	5	1	1–104	0.016

* The caption explains the significance of the asterisk

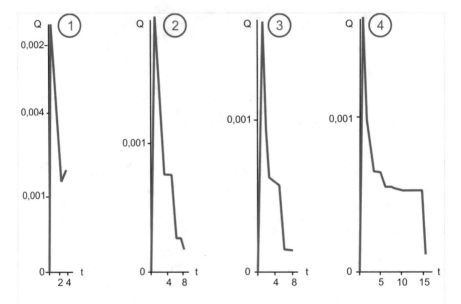

Fig. 6.8 Unit hydrographs obtained in the four tests

The discharge of the Simat spring started being monitored by the IGME in February 1973 as part of the Lower and Middle Jucar Hydrogeological Research Project. In Simat de Valldigna, there is a rainfall station whose measurements are considered representative of the rainfall in the catchment area, although its average altitude is greater than that of Simat. This translates into a deficient reconstruction of the simulated series, especially in the lowlands. Floods are not well reproduced, either.

That is why in the following the memory effect was increased to eight months, a maximum was imposed on the impulse response in Month 1 and the calibration period of the impulse response was increased from 19 to 43 months. With four calculation phases, the impulse response is not completely annulled after eight months, although the average norm improves

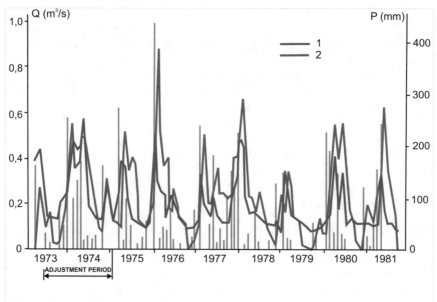

Fig. 6.9 Results of the first pass (input: rainfall)

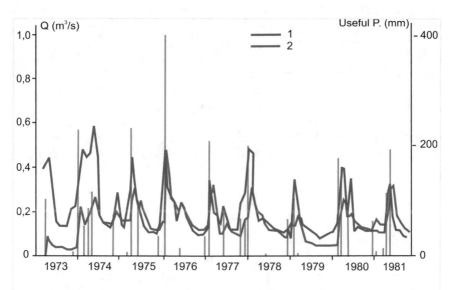

Fig. 6.10 Results of the fourth pass (input: useful rainfall)

significantly with respect to the first pass. The improvement in the adjustment is also visible in Fig. 6.10, although the lows are still not reproduced completely. Everything seems to indicate that the system has an even longer memory effect.

If, instead of identifying the impulse response at 43 months, the whole period is used for calibration, the standard improves appreciably, although zero is not reached in the impulse response either (Fig. 6.7).

Finally, in order to bring the calculation closer to the physical phenomenon, a final pass was carried out using as the system input not gross rainfall but useful rainfall. To obtain this, we deduct actual evapotranspiration —calculated by Thornthwaite's method—from the rainfall measured at Simat. In addition, the memory effect was increased to 16 months, obtaining a value close to zero after 15 months (Fig. 6.7) in a surprising result. With these data, it was observed that the adjustment in the low-water level improved significantly, although the floods reproduced rather worse, the resulting norm being higher than the one obtained in the previous three.

Torcal de Antequera with daily data

A total of six valid tests were conducted, and the characteristics of which are summarized in Table 6.3. In the first three, raw rainfall was used as input, while in the remaining three, it was the useful rain, estimated by applying the Thornthwaite method to the daily data.

In the first test, the data were introduced without restriction regarding the impulse response (Fig. 6.11) but the memory effect was set at 128 days. As a consequence of this and by not imposing smoothing, a response was obtained with a multitude of small steps, although the norm in the last calculation phase (six calculation phases, in this case) was not excessively high. The results are shown in Fig. 6.12.

From Fig. 6.11, which reproduces the 'observed' and calculated hydrographs, it can be deduced that there is a notable divergence from the pronounced lows, as well as from the great flooding of January to March in 1977. For this reason, we considered that it was necessary to fix a maximum in the impulse response function and to prolong its memory effect to try to obtain a

Table 6.3 Summary of the various tests

No.	Impulse response (days)	Calibrated period (days)	UFC (LCP)	Maximum (days)	Simulated period (days)	Mean CFU standard (days^2)
1	128	129–620	6	without	1–1100	0.0394
2	256	257–620	5	30	1–1100	0.0581
3	256	257–100	5	30	1–1100	0.0156
4[*]	256	257–100	5	30	1–1100	0.0179
5[*]	256	257–620	5	30	1–1100	0.0829
6[*]	128	129–620	5	16	1–1100	0.0497

[*]useful rain is used as input; UFC last calculation phase

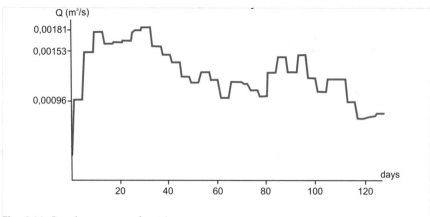

Fig. 6.11 Impulse response of test 1

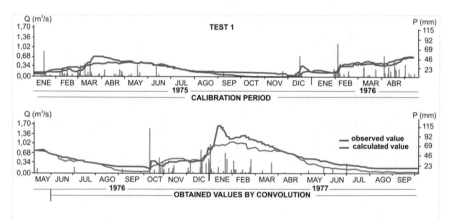

Fig. 6.12 Results of the first test

greater adjustment in the low-water levels. We set 30 and 256 days, respectively.

The norm did not improve but increased slightly, after the second pass. However, by using the entire length of the register as the calibration period, a noticeable improvement was achieved. The impulse response (Fig. 6.13) could be extended up to 256 days, although it was practically cancelled by 190 days. Although the adjustment improved appreciably, the disparity persisted in both low-water values and in major floods, as described above (Fig. 6.11).

To achieve a better adjustment and to identify an impulse response more in line with the physical reality of the phenomenon, we decided to use useful rainfall as an input, instead of raw rainfall. In the three tests (Table 6.3), the lowest standard deviation was obtained using the whole period as the

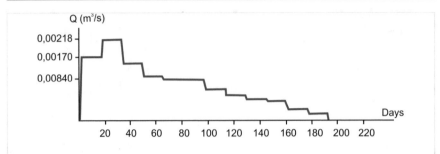

Fig. 6.13 Impulse response of test 3

calibration interval. If all the remaining data were maintained and calibration was only undertaken only with the first 620 days, the value of the standard increased significantly. In both cases, 256 days was used both as a memory effect value and a maximum in the impulse response at 30 days; however, the impulse response was practically over by 161 days. It was found that extreme values were not adequately reproduced.

With regard to the operation of the aquifer, the results obtained reveal its great inertia, with a memory effect of nearly 200 days—an exceptional value. It should be noted that these same data were processed by means of simple and cross-correlation analyses in Chap. 4 [15], obtaining on that occasion a value of the order of 70 days. Regardless of the influence from measurement errors, the most satisfactory explanation is that, in deconvolution, exhaustion has considerable weight when reproducing the deconvolutionary series, and in correlation analysis, the greatest weight is in the effect of the flood impulses, in this case.

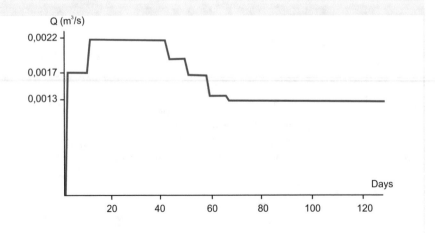

Fig. 6.14 Impulse response of test 6

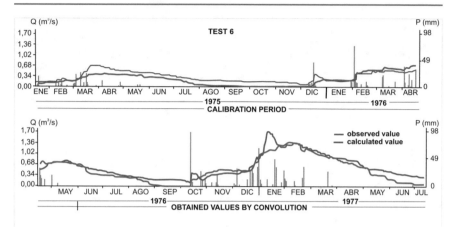

Fig. 6.15   Results of test 6

A final attempt was made to reduce the memory effect to 128 days and the maximum to 16 days, with the idea of obtaining a better adjustment of the floods. The result was not as good as desired, but it reduced the norm on the fifth pass. The most interesting aspect to highlight is that the impulse response after 128 days maintains a high value (Fig. 6.14), indicating that the system has greater memory. This translates, in convolution, to the presence of greater divergence in pronounced lows, despite achieving a relatively greater adjustment of the maximum values (Fig. 6.15).

In this sense, the program's failure to reproducing the long-sustained lows could be regarded as due to the presence of a slow lateral (or vertical ascending?) potential feed that contributes a flow with very minor variation over time. Deconvolution cannot, in this case, represent this contribution. A new analysis could be carried out to discount a constant amount of low-water flow and deconvolve the residue, so that the model would be in this case $S(t) = \text{constant} + E * \Phi$.

On the other hand, the difficulty of adjustment and reproduction of certain floods, especially that of January 1977, is regarded as being due to the non-stationary nature of the system. The heavy rainfall of December 1976 and January 1977 affects the flow regime (start of operation of certain ducts?) in such a way that they modify the physical system; therefore, the modelling is unable to reproduce it, since it requires a stationary nature.

The contrast is even more pronounced if the results are compared with those obtained when applying the procedure to the monthly data from Torcal [15, 16]. In this case, a memory effect of more than a year's duration was obtained, which is highly paradoxical. Nevertheless, good adjustments were obtained with memory effects of eight months. In this case, we concluded that, although the methodology covers diverse objectives (reproduction of the hydrograph, generation of data, completion of lagoons, and so on), the physical result was not consistent with the reality. In addition, the physical significance of an average monthly flow has very little to do with an instantaneous flow or an average daily flow, so the results obtained are not comparable.

Finally, several practical considerations emerge from the above. Torcal de Antequera constitutes an aquifer system of great inertia, long memory effect and good regulation, whose hydrogeological behaviour is close to that of a porous medium and therefore possibly modifiable by methods applied to that medium. The probability of a successful borehole in this catchment, therefore, must be very high, and this is corroborated in reality since there are several high-yield shafts next to the main spring.

Tank models: Application to Torcal de Antequera

We applied the procedure described to the karst system of Torcal de Antequera, La Villa spring and the daily precipitation and temperature records of a representative station in the sector over six years. As the input data, the daily rainfall was calculated using the EVADIA program [17]. The first four years were used as a period of calibration for the model and the last two to check the goodness of the fit of the procedure. A weekly time interval was used.

The Maillet formula can be adjusted to the decrement hydrograph in Fig. 6.15, but this is not always possible, in which case another function will be used. To calculate the volume that is contributed to the saturated band due to precipitation, the procedure is as follows [18]:

$$Q_i = Q_0^{e-\alpha_t} \text{ or } Q_i = Q_i^{e-\alpha\Delta_t}$$

where: Q_0 = flow at the beginning of the decreasing; Q_i = flow at instant i; α = base flow coefficient; and Δ_t = time interval between two instants. If we assume Δ constant t, we have: $e^{-\alpha\Delta t} = B$ = constant. The volume contributed by the precipitation at instant t that is susceptible to drainage by the spring is given by:

$$VA_i = \int_i^\infty Q_i e^{-\alpha t} dt \int_i^\infty B Q_{i-1} e^{-\alpha t} dt = \frac{Q_i - BQ_{i-1}}{\alpha} \qquad (6.19)$$

From a historical series of n flow values, volumes stored in the saturated zone due to external inputs, generally rainfall or return of irrigation, can be calculated using function in Chap. 1 [1] $n - 1$, in the case of a closed aquifer system. We assume that the unsaturated zone exerts a weighting function on the effective infiltration, decomposing it into a set of volumes that contribute to the aquifer over the time following the instant in which it was produced. We also consider that this decomposition is linear in time; that is, its

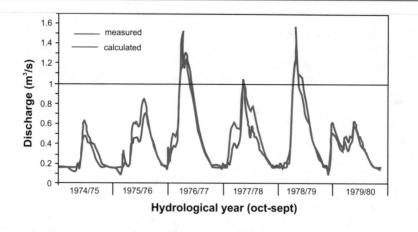

Fig. 6.16 Actual and simulated hydrographs obtained weekly at the Torcal de Antequera aquifer

characteristics do not depend on time and there is a proportionality between effective infiltration and transmitted volumes, so that this regulation can be expressed through an integral of convolution:

$$VI_i = \int\limits_0^{+\infty} VA_{i+t}\lambda_t \mathrm{d}t$$

and, using Hankel's transformation [19]:

$$VA_i = \int\limits_{-\infty}^0 VI_{i-t}\lambda_t \mathrm{d}t_2$$

Figure 6.16 shows the theoretical hydrograph obtained by the above method, compared to the real hydrograph. As mentioned above, the first four years correspond to the period of adjustment of the parameters intervening in the model, and the last two correspond to the simulated period.

6.4.4   Box 2: Application of ARMA Models

6.4.4.1   Methodology and Physical Justification

Univariate models are based on analysis of the stochastic structure of a time series, with the aim of making short-term forecasts. Complete data, or a generated synthetic series that preserves the average statistics, especially the time correlation (autocorrelation function) of the underlying process in the original series, are available. In this example, we want to demonstrate the applicability of ARMA models to the study of time series of springs that drain karst aquifers. It is a question of providing an explanation for the physical results of the autoregressive parameters and moving averages of an ARMA model applied to karst springs, specifically Aliou spring in the French Pyrenees, characterized by quick flow through a karst network (in Chap. 1 [18], in Chap. 4 [13] and [20]).

The self-regressive model in applied moving averages, ARMA (p, q) of constant parameter, has the general expression in Chap. 5 [2]:

$$z_t = \sum_{i=1}^{p} \emptyset_i \, z_{t-i} + a_t - \sum_{j=1}^{q} \theta_j \, a_{t-j} \qquad (6.20)$$

where z_t represents a standard periodic time series, and ϕ is autoregressive θ coefficients varying in time and moving averages, respectively, and a is independent and normally distributed variable (white noise).

To investigate the physical sense of an ARMA model applied to the flow of a karst spring, we use the graph in Fig. 6.4, in which the variables represent discrete values. We assume that the functions related to volume transfer are linear. Precipitation at a given instant P_t is distributed between infiltration into the saturated zone that flows slowly (base flow), aP_t; evapotranspiration bP_t; and rapid circulation into the conduits (fast flow), which in an isolated system would be equal to $(1-a-b)\, P_t = dP_t$. Parameters a, b and d represent the various fractions into which precipitation is divided. The discharge of the spring at instant t is given by the expression $Q_t = cV_{t-1} + dP_t$ where $c\, V_{t-1}$ indicates the fraction of the volume stored in the previous instant drained by the spring. The volume stored at instant t is equal to $V_t = V_{t-1} + a\, P_t - c\, V_{t-1} = (1 - c)\, V_{t-1} + A\, P_t$. Combining these two equations with those previously obtained for Q_{t-1}, V_{t-1} and V_{t-2}, gives:

$$Qt = (1\,c-)Q_{t-1} + dP_t[d\,(1-c) - ca]P_{t-1} \qquad (6.21)$$

This expression is similar to that of an ARMA model $(1, 1)$, in which precipitation is a random and independent variable; $(1 - c)$ equals Φ and indicates how the saturated zone drains through small discontinuities (in diffuse flow); dP_t equals the random term a, and represents the fraction of the precipitation that circulates essentially through large conduits at instant t; and $[d\,(1-c) - ca]$ equals θ_1. This last term has no clear physical meaning, although it may be the fraction of the precipitation that circulates through large conduits in the previous instant, so it could be considered representative of the degree of organization of the karst network in the system, likely to vary by precipitation regime.

The procedure followed here, in the first place, was for the series of flows to be transformed to reduce the bias and eliminate their periodicity. The next step was to calculate autoregressive parameters and moving averages using the least squares method, solving the system of equations with the Marquardt algorithm (1963). Subsequently, the ARMA model was found to have the best fit to the available series and, finally, the goodness of the model fit was demonstrated.

6.4.4.2   Transformation of Series

The period studied (1 October 1970–30 September 1975) covers five hydrological years of daily data (1825 data items). The flow hydrograph shows that storage is practically nil, with a clear periodicity due to the intense peaks in periods of high water related to the distribution of precipitation across the sector in Chap. 4 [8, 9]. This periodicity must be eliminated by transforming the original series by applying a constant parameter ARMA model.

The mean statistics of the data series are: $m = 0.47$ (mean), $s = 0.94$ (standard deviation) and $g = 6.36$ (high bias values indicate lack of normality in the series). To reduce the bias, after attempting several transformations, a logarithmic transformation was adopted as the most appropriate. The value of m of the new series, $y_t = \log(q_t)$, is $0-0.85$, s is 0.71 and g is 0.02; it is noted that the new value of g is much lower.

To analyse the periodicity of the most significant statistics (mean and standard deviation), we grouped the terms of the series and t into their corresponding years. This new series was represented by $y_{V,T}$, where V indicates the years and T the number of intervals into which the year is divided. Periodicity is eliminated by transformation $z_{v,T} = \frac{y_{V,T} - m_T}{s_T}$

To simplify the process, we calculated the m_T and s_T by means of the most significant coefficients, obtained with the Fourier transform of both statistics.

Figure 6.17 confirms that the periodicity has been eliminated, since many of the values of $R_{1,T}$ are close to 0.8. Consequently, an adjustment to an ARMA model of constant parameters is possible. The m value of the z_t series (obtained from $z_{V,T}$) is -0.02, s is 1.13 and g is 0.03.

6.4.4.3   Adjustment and Goodness of Model Fit

The adjustment of the models was undertaken for the five years of z_T. To obtain the autoregressive parameters and the moving averages that the ARMA model must have, we calculated the partial autocorrelogram of z_t with a 95% confidence limit (Fig. 6.18). A powerful first autoregressive parameter is obtained; consequently, the ARM model to be adjusted has a single autoregressive parameter. To set the number of parameters in moving averages, five attempts were made (Table 6.4). Of all the models analysed, the ARMA (1, 2) has the lowest value of Sa^2. Thus, the definitive ARMA model (1,2) is expressed as:

Fig. 6.17 First
autocorrelation coefficient
(R_1, T) of Aliou's zV, T series

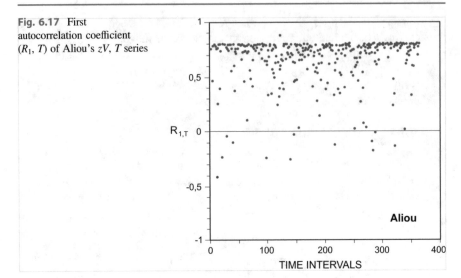

Fig. 6.18 Partial
autocorrelogram of the z_t
series of Aliou flows

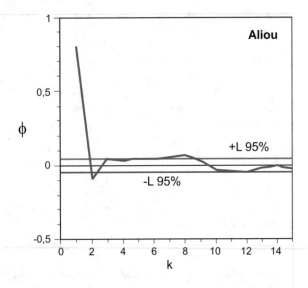

$$\hat{z}_t = 0.811\,\hat{z}_{t-1} + a_t + 0.071 a_{t-1} - 0.119 a_{t-2} - 0.003 \qquad (6.22)$$

A procedure to calculate the goodness of the model fit, adjusted in each case, involves testing the independence of the residuals. The statistic $Q = 16.3$ for $L = 20$ is that the value of $X^2 = 27.6$ with $L - p - q = 17$ degrees of freedom and with a level of significance of 0.05. Consequently, the hypothesis of waste independence is accepted.

Table 6.4 Parameters of the models adjusted to Aliou z_t series: C, constant; $\Phi\rho$ autoregressive coefficients; θ_q, coefficients of moving averages; Sa_2, sum of squares residues; Q, value of X_2 for the first 20 coefficients of autocorrelation of residues; and AIC, the Akaike information criterion (1974) of the parameters of parsimony

Model	C	Φ_1	Φ_2	θ_1	θ_2	θ_3	Sa^2	Q	AIC
ARMA (1,0) ...	0.001	0.798	–	–	–	–	848.4	46.4	−1.396
ARMA (1,1) ...	−0.002	0.744	–	0.149	–	–	839.2	27.3	−1.414
ARMA (1,2) ...	−0.003	0.811	–	−0.071	0.119	–	832.9	16.3	−1.425
ARMA (1,3) ...	−0.006	0.822	–	−0.049	−0.119	−0.008	832.4	25.5	−1.424
ARMA (2,0) ...	−0.011	0.862	−0.077	–	–	–	840.4	25.5	−1.411

Confirmed by the hypothesis of the independence of the residues, we fixed the synthetic series $\hat{z}$. The random values were generated with the same log-normal distribution ($\beta = 1.6$, $\mu = 0.49$ and $\sigma_m = 0.36$), mean and standard deviation of the residues obtained from the ARMA model adjustment (1.2). To obtain the series of synthetic spring flows, we followed the reverse process, transforming the series. Once the values of $\hat{z}_t$ were grouped by years, the $\hat{z}_{VT}$ were transformed by the expressions:

$$\hat{y}_{V,T} = \hat{s}_T \hat{z}_{V,T} + \hat{m}_T \quad \hat{q}_{V,T} = 10^{\hat{y}_{V,T}} \tag{6.23}$$

The synthetic hydrograph (Fig. 6.19) has a morphology very similar to the real one, with sharp, random peaks and a very low storage power system. The correlograms of the q_t and $\hat{q}_t$ series (Fig. 6.20) are very similar, indicating that the main statistics are retained in the adjusted model.

All the types of models seen in this section have their main application in studies of the rainfall relationship, or eventually in the piezometric—rainfall domain—of an aquifer [21], with three most important objectives [22]:

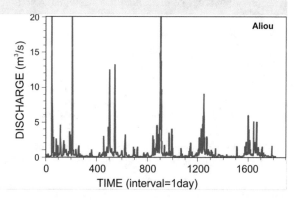

Fig. 6.19 Synthetic hydrograph obtained with the model adjusted to Aliou flows

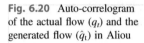

Fig. 6.20 Auto-correlogram of the actual flow (q_t) and the generated flow ($\hat{q}_t$) in Aliou

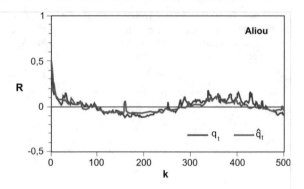

- Reconstruction of historical series of flows from a karst spring, when there is a short period of registration (at least five years) of known precipitation in the basin.
- Forecasting the base flows by means of hypotheses about expected rainfall, which in our latitudes, for the months of April and May, would be zero.
- Calculation the hydrograph in the natural regime and comparison to the real one, to determine the influence of pumping from the area surrounding aquifers that start to be exploited after the adjustment period.

To these three objectives, we could add another two:

Estimation of missing data in a series of a spring's flow (also applicable to surface basins).

Evaluation of the regulatory power of the unsaturated zone, so little known at present, by means of appropriate modification of the models.

The practical interest of this type of model is very restricted, since the entire aquifer is synthesized in a black box, thus the parameters involved have no real physical significance that is evident. Neither is any account taken of the dimensional or hydrodynamic parameters of the aquifer, nor its spatial distribution. Although their interpretation is not simple, the models can provide qualitative information (especially of relative value) on the karst's structure.

6.4.5   Box 3: Application of Traditional Models

Torcal de Antequera, northeast of the province of Malaga [23], and the Sierra Grossa aquifer in Valencia are the two examples already examined in Chap. 3 [12].

6.4.5.1   Torcal de Antequera

Description of the model It was deduced from the application of correlation and spectral analysis to the series of flows of springs and the rainfall of these two massifs in Chap. 4 [13] that these systems could be simulated by conventional methods (Photos 6.1 and 6.2). There is a long memory effect, strong inertia and good response times. The study of yields of boreholes in Chap. 3 [13] into the intergranular porous medium had already pointed to the presence of an increasing 'equivalent transmissivity' from the recharge to the discharge areas. All of this was taken into account in establishing the conceptual model, as well as the presence of highly transmissive drains and capacitive blocks in Chap. 3 [8, 9] and in Chap. 4 [6]. A simulation code in finite differences was used [23].

Figure 6.21 shows the chosen net size, a grid of variable size with lengths between 1000 and 100 m. In this configuration, there are two large conduits or drains. The size of the net was reduced with increasing proximity to the spring to lend more sensitivity to this sector, in respect the considerations already made. The limits of the aquifer are clear, as well as its impermeable nature in all cases.

Photo. 6.1 View of the Torcal de Antequera from the visitors centre (*Photo* A. Pulido)

Photo. 6.2 Three of the boreholes near the spring of La Villa, one now decommissioned (*Photo* A. Pulido)

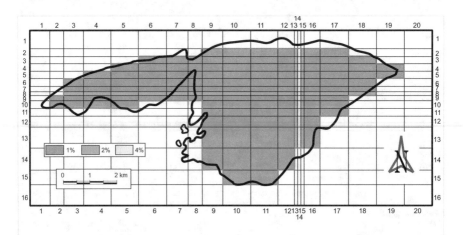

Fig. 6.21 Grid and initial values of the storage coefficient (S_0)

The initial data of S and T are shown in Figs. 6.21 and 6.22. The values of S were considered homogeneous throughout the aquifer (1%), apart from the drains, which used 4% for north–south and 2% for east–west. The original permeability values were highly variable, with a progressive decrease with distance from the spring in order to stage an equivalent increase in transmissivity with proximity to the discharge area. Along with the spring, values of 60 m/d were adopted, decreasing to 2 m/d in the most distant sectors: in the north–south drains, 100 m/d; and in the east–west drains, 60 m/d. The bottom of the aquifer was assigned a height of 500 m, because the real values are unknown.

The initial hydraulic head (Ho) included in Fig. 6.23; it varies between 600 m at La Villa spring and a little over 620 m at the western end. To determine it, the strategy was devised of adopting the values that best reproduced the springs' flow in the initial interval (base flow period) at the

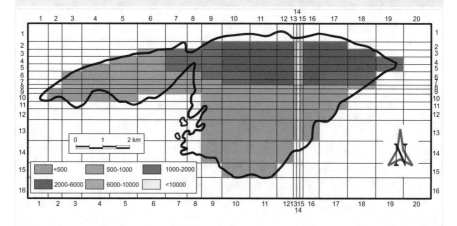

Fig. 6.22 Initial transmissivity values

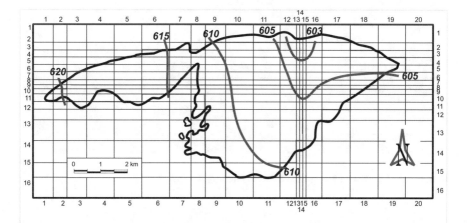

Fig. 6.23 Initial hydraulic head values

time of adjustment. Infiltration was estimated using rainfall data from Antequera and Torcal itself (from February 1975) and arachidonic temperatures. Useful rainfall was estimated as the difference between P and ETR, calculated using the Thornthwaite method, for a water reserve usable by plants of 100 mm [17].

The adjustment criteria were the flows measured and simulated at La Villa spring (Photo 6.3) and the overall budget. The simulation of the spring condition was undertaken by introducing an output transmissivity factor large enough to guarantee the constancy of the emergency elevation yet avoiding introducing error conditions into the program: the value adopted was 5.E5.

Results obtained From the initial data entered, and on the basis of the conceptual models cited, the simulated response to the measured hydrograph was adjusted. The peaks did not reproduce well and the low values remain above those measured. The value of S in the north–south drain was reduced to 2% without noticeable improvement to the result. At Test 3, the value of S was significantly reduced (1% in the east–west drain, and 0.7% in the north–south, a value also assigned to the rest of the aquifer). With these data, the result improved appreciably, although the weighted mean value of S seemed too low. The following modifications were applied to T, little marked in Test

Photo. 6.3 La Villa spring in March 2014 (*Photo* A. Pulido)

4 and much more abrupt in Test 5, going from about 1500–2400 m²/d, with a notable improvement in the result. The weak point was the initial flows due to the high initial hydraulic head values that were introduced.

At Test 8, the result obtained in the pumping test in one of the supply boreholes near the spring was taken into consideration, with 60,000 m²/d for T and 10% for S (Figs. 6.24 and 6.25). The adjustment obtained an improvement, although a disparity remained in the initial flow. Hence, there was a final attempt to modify the Ho values to obtain the hydrograph of Fig. 6.26.

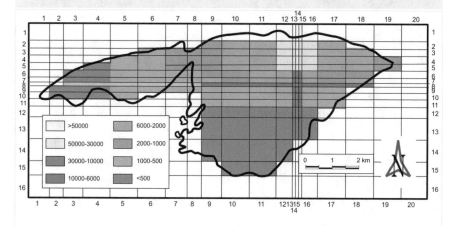

Fig. 6.24 New transmissivity values

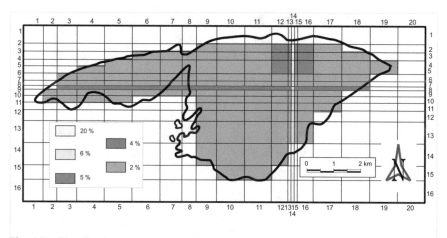

Fig. 6.25 New S values

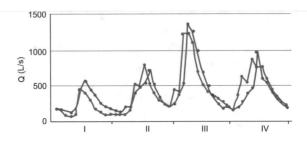

Fig. 6.26 Real and simulated (red) hydrographs obtained

The average data from this test were considered acceptable, with $S = 1.9\%$ and $T = 2\ 800\ \text{m}^2/\text{d}$, with a fairly accurate water budget and a mean gradient in the system of 0.25%, which is not realistic for a karst medium. Several new modifications in T and S allowed a better adjustment of the simulated response, reaching a final average gradient of 0.8%, S of 1.13% and T of the order of 800 m^2/d.

It follows that certain karst aquifers can be simulated by methods normally used to simulate the intergranular porosity medium, with fairly acceptable results. Another aspect of interest in the simulation results is related to the level fluctuation values obtained in the simulation, with rises exceeding 30 m, as measured in real karst aquifers. One result is that the value of the average storage coefficient obtained is less than 2%.

6.4.5.2 Sierra Grossa Aquifer (Valencia)

Geological framework
The aquifer of Sierra Grossa in the strict sense has an area of 350 km^2, from Fuente la Higuera to Simat-Jeresa-Gandía, with an elongated shape in the ENE direction of 60 km long and 15 km wide in Chap. 3 [12]; Fig. 6.27). The simulations were carried out at the Centre d'Informatique Géologique de l'Ecole des Mines de Paris, Fontainebleau.

The aquifer materials with the greatest development—limestone and dolomite —(310 km^2) belong to the Creu formation (Upper Cretaceous) and are over 600 m thick. They follow those of the Jaraco (Middle Cretaceous; 22 km^2) and Infierno (Upper Jurassic) formations, which occur at the eastern end. There are also two poljes [24], one open (Maxuquera) and the other closed (Barig). The aquifer is well defined at the edges, which are all impermeable (Miocene marl and/or Triassic clay with gypsum). Only the eastern edge has permeable detrital quaternary materials, and there is a small sector to the south of Jativa with material of the quaternary of Canals.

The structure of Sierra Grossa is relatively complex and has markedly different features from its extreme west to the east. The anticlines have a clear

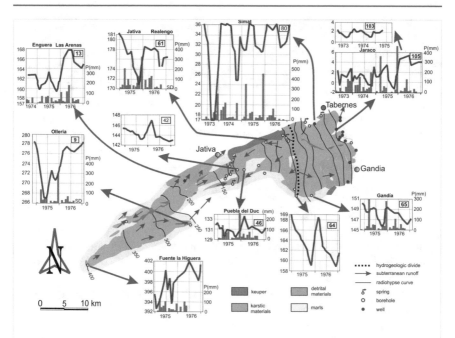

Fig. 6.27 Sierra Grossa unit and piezometric evolutions in Chapt. 2 [15]

asymmetry, with the southern flank gently sloping and the northern vertical flank often inverted or even overthrusting. In the eastern extremity, there are dome structures that Champetier [25] assigns to a 'zone of interference between Prebetic and Iberic'.

With regard to the operation of the aquifer unit, it should be noted that the main supply is from direct infiltration of rainwater, and to a much lesser extent from losses from rivers. The main exits are through springs, the most important being those along the River Albaida and those at the eastern end. Along the eastern edge, there are hidden losses towards the detrital aquifer of the Gandía-Denia plain. Finally, an important aspect of the water outlet is pumping.

Simulation

The deterministic model used in the simulation aimed to account for the peculiarities of the more or less karstified carbonate medium. Indeed, a possible conceptual model involves the presence of capacitive blocks of low permeability and storage, separated by discontinuities (fractures and/or conduits) that are highly transmissive (the transmissive element). This model should be superimposed on the effect, generally resulting from karstification, of a 'hierarchization', in the sense of increasing the equivalent transmissivity from the recharge areas to the discharging areas in Chap. 3 [13], as deduced from analysis of the yields of the catchments in the karst environment.

NEWSAM software, designed at the Centre d'Informatique de l'Ecole des Mines in Paris, was used for the simulation. It permits quantitative simulation of the transport of water and matter in multi-layered aquifers, in both a permanent and a transitory regime [26]. It is a finite difference model that allows the discretization of the aquifer in square nets of variable dimensions, which gives it a versatility that other simulation programs do not have: it caters for up to four net sizes. This has been used advantageously to provide greater precision to calculations in certain sectors that have more data (discharge areas, essentially) yet to leave wider net sizes in recharge areas.

Sierra Grossa has been simulated as a monolayer aquifer, as shown in Fig. 6.28. A total of 629 nets were used, varying between 4 and 0.0625 km² across. Also included in the area to be simulated are the sectors in which carbonate materials are covered by detrital materials (poljes and part of the eastern edge on the boundary with the Gandía-Denia plain), so that the total simulated surface area measures 416 km².

Smaller nets were used in the discharge areas, specifically along the River Albaida (main discharge) and next to the Simat de Valldigna spring. In addition, they were used in a central strip created between the western end and the River Albaida. Larger net sizes were placed to coincide with the

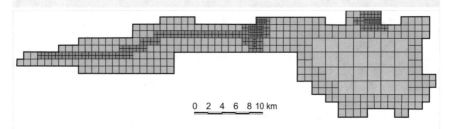

0 2 4 6 8 10 km

Fig. 6.28 Grid used in the discretization of the aquifer

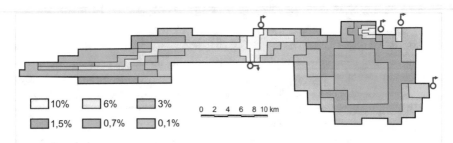

☐ 10%   ☐ 6%   ☐ 3%   0 2 4 6 8 10 km
☐ 1,5% ☐ 0,7% ☐ 0,1%

Fig. 6.29 Initial storage coefficient values

wider area of the aquifer (Cuatretonda-Barig-Almiserat) that had practically no data at all.

Figures 6.29 and 6.30 show the initial values of the storage coefficient (S_o) and transmissivity (T_o). In both cases, an attempt was made to consider the available data as well as what has been stated in relation to the conceptual model. The highest values of S_o were found next to the two spring areas (Simat and Albaida), with 10%. Bordering these areas and along the western 'drain', there were values of 6%, and these were surrounded in turn by meshes with values of 3 and 1.5%. In the recharge areas, which are intended to represent the average storage coefficient of the blocks, values of 0.7 and 0.1% were found.

Top values varied by more than two orders of magnitude between the extremes, with 50,000 m^2/day in the spring areas, 20,000 m^2/day in surrounding areas and in the main 'drain', 10,000 m^2/day on the eastern edge, discharge, down to 500–200 m^2/day in the recharge areas. It was attempted to simulate the presence of the 'hierarchization' derived from karstification.

The initial hydraulic heads (H_o) used were as shown in Fig. 6.27, corresponding to medium-type water table contours. Most of the nets along the River Albaida were regarded as discharge with an imposed hydraulic head, as well as those corresponding to the Simat de Valldigna spring in and the two easternmost nets bordering the Gandía-Denia plain. The unitary recharge introduced varied from one net to another, in the same way that the rain falling on the basin varies. With the exception of the permeable eastern edge, they were considered to be impervious limits of null water exchange.

With the above-mentioned input data, the aquifer was simulated first under a permanent regime. Following nine tests in which the values of T and S were adjusted, the piezometric surface was reproduced quite faithfully. The data from T (Fig. 6.31), S and H thus obtained were used as starting values in the simulation under a transient regime. The simulated period was October 1974 to September 1976; that is, two hydrological years. The adjustment criteria were:

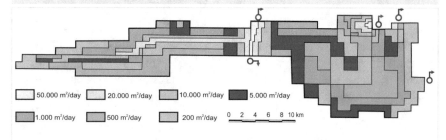

| | 50.000 m²/day | | 20.000 m²/day | | 10.000 m²/day | | 5.000 m²/day |

| | 1.000 m²/day | | 500 m²/day | | 200 m²/day | 0 2 4 6 8 10 km |

Fig. 6.30 Initial values of transmissivity

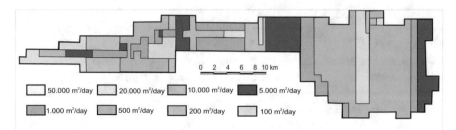

Fig. 6.31 Transmissivity values after adjustment under a permanent regime

- Those nets where piezometric measurements were available (11 piezometers in total; Fig. 6.27).
- The unit's two main sources' measured and simulated flows: discharge into the River Albaida and the Simat de Valldigna spring.

In each of the nets, the infiltration of rain and the pumping outlets were calculated. The former was taken from the rainfall at the Simat de Valldigna, Jativa and Fuente la Higuera stations, minus actual evapotranspiration calculated by Thornthwaite's method for a water reserve usable by plants of 50 mm. The values obtained in each of the stations were considered representative of the eastern, central and western third, respectively.

The pumped volumes were known roughly from the IGME water point inventory data. During the entire simulated period, less than 50 hm^3 were extracted, while the rain recharge slightly exceeded 250 hm^3.

After a total of seven tests, the simulation results were considered to be satisfactory. The final values of S are shown in Fig. 6.32, together with the piezometric evolutions in seven nets, obtained through simulation and field measurements. At the end of the simulation period, an increase of about 6 hm^3 of water stored in the aquifer is obtained, which is reflected in the final state of the piezometry of some areas. Figure 6.33 shows the evolution of the flows drained by the River Albaida and by the Fuente Mayor de Simat de Valldigna spring in the seventh test.

The results obtained through this simulation are quite satisfactory and show the feasibility of using this tool in the simulation of more or less karstified carbonate aquifers. Earlier identification of the extent of karstification by means of correlation and spectral analysis or deconvolution, for example, allows us to have an idea of the goodness of fit of such an approximation.

This simulation constitutes a more advanced stage in the approximation in the simulation of Mediterranean karst aquifers than that previously carried out for the aquifers of Torcal de Antequera [23, 27] and Simat de Valldigna. In fact, in both cases, the reproduction of the flows measured at the main spring was used as an adjustment criterion, without reliable data on the piezometric

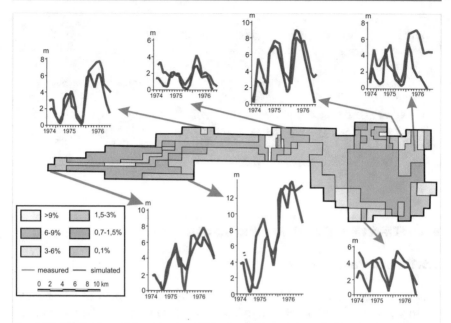

Fig. 6.32 Final values of the storage coefficient and piezometric evolutions across seven nets

Fig. 6.33 Flow rates measured (1) and obtained by simulation (2) in River Albaida and Fuente Mayor de Simat de Valldigna spring

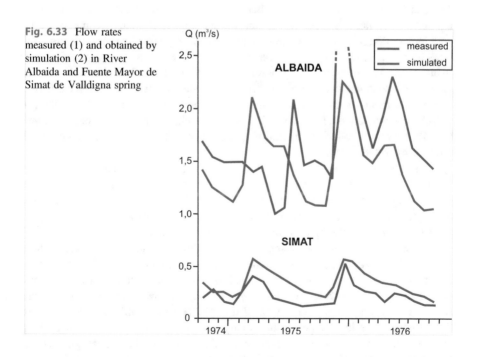

levels. The presence of piezometric evolution data at some points of Sierra Grossa aquifer gives a certain guarantee to the final distribution of the T and S values obtained.

The two aspects attempted to be reproduced in the simulation process were the presence of highly transmissive sectors next to most of the system with much lower values (several orders of magnitude), and a progressive increase in the 'transmissivity' of the aquifer from the recharge areas to the discharge areas.

6.5 'Traditional' and Specific Models

6.5.1 General Aspects

This is the case for the models normally used in the simulation of intergranular porosity aquifers. In this respect, correlation and spectral analysis, both simple and crossed, applied to the rainfall of the basin and to the discharge values of the spring can serve as a starting element to determine the *feasibility* of this type of modelling. The fact is that, in general, the number of real data referring to the parameters of aquifer in each net is scarce. This is abrupt topography, and the catchments are in very localized sectors. Moreover, in many cases the adjustment criterion is the flow measured in a spring, making some authors propose calling this type of model 'white box'. All fall within the generic denomination of distributed models and, in the case of karst, can become extremely detailed and verifiable in the very difficult real environment. A classification may be: Discrete fracture network (DFN) approach; discrete channel network (DCN) approach; equivalent porous medium (EPM); double continuous medium (DC) approach; and discrete-continuous (DC) or hybrid combination [28].

There are many examples of the application of traditional models to the simulation of karst aquifers. One of the first was that of the limestone aquifer of the Cenomanian-Turonian of Israel [29], totalling 3500 km^2 in extent, obtaining the adjustment for values of T between 5000 and 100,000 m^2/day and from 4.2 to 6.3% for S (0.3% in confined sectors) by means of a R–C model.

One case is the Vaucluse spring basin [30], another is the multi-layered Oued R'hiv Valley in Algeria [31], where one of the simulated aquifers is limestone. The limestone of the Caramy river basin (South of France) was simulated in the same way, using a coupled model (*modèle couplé* [32]) that allows the simulation of both surface run-off and underground flow of a basin. The use of this type of model in the simulation of karst aquifers has many practical advantages over black-box models, since it takes into account the spatial distribution of the aquifer parameters, and allows future exploitation and the expected evolution of levels and spring flow to be simulated. However, if we consider the conceptual model of the karst aquifer of large blocks of low transmissivity—the *capacitive element*—separated by highly transmissive conduits—the *transmissive element* (in Chap. 3 [8, 13]), there is always a doubt about the real representativeness of the values of T and S of each net that allow adjustment of the model. If a borehole is drilled in a determined net, would we obtain the values of T and S that allowed the adjustment of the model, or the opposite: is it possible to find values between the T of the conduit and the T of the block?

The 'specific' models are based on the conceptual model of the previously synthesized karst aquifer; that is, blocks and conduits. Blocks are assigned a permeability of the order of 10^{-5} m/s (the average permeability obtained in pumping tests in existing boreholes), separated by the permeability of *conduits* between 10^{-1} and 10 m/s [33, 34]. From this type of model, and by playing with the dimension of the blocks, it is possible to reproduce the hydrographs of karst springs [35] with very good approximation. However, while traditional models raise the question of the actual physical significance of such a simulation, in this type of model, it is evident that, although the form of the hydrographs is reproduced with great fidelity, the simulated grid's resemblance to the real aquifer may be practically non-existent.

6.6 Further Reading

Bailly-Comte, V., Borrell-Estupina, V., Jourde, H., Pistre, S. 2012. A conceptual semi-distributed model of the Coulazou River as a tool for assessing surface water—karst groundwater interactions during flood in Mediterranean ephemeral rivers. *Water Resources Research,* 48: https://doi.org/10.1029/2010wr010072.

Dörfliger, N., Fleury, P., Ladouche, B. 2009. Inverse modeling approach to allogenic karst system characterization. *Ground Water,* 47 (3): 414–426.

Hartmann, A., Goldscheider, N., Wagener, T., Lange, J., Weiler, M. (2014) Karst water resources in a changing world: Review of hydrological modeling approaches. *Reviews of Geophysics,* online.

Jeannin, P-Y., Malard, A. 2018. *Visual KARSYS* (https://visualkarsys.isska.ch/). Swiss Institute for Speleology and Karst-Studies.

Ladouche, B., Marechal, J., Dörfliger, N. 2014. Semi-distributed lumped model of a karst system under active management. *Journal of Hydrology,* 509: 215–230.

Malard, A., Jeannin, P-Y., Vouillamoz, J., Weber, E. 2015. An integrated approach for catchment delineation and conduit-network modeling in karst aquifers: Application to a site in the Swiss tabular Jura. *Hydrogeology Journal.* https://doi.org/10.1007/s10040-015-1287-5

Mero, F. 1978. *The MMO8 hydrometeorological Simulation System. Basic concepts and operators guide.* Tel Aviv: Rapport Tahal T/78-02.

Pardo-Igúzqiza, E. 2019. A review of fractals in karst. *Int. J. Spel.,* 48 (1).

Watson, G.N. 1966. *A Treatise on the Theory of Bessel functions.* Cambridge Univ. Press, 804 p.

Yevjevich, V. 1972. Stochastic processes in hydrology. *Water Resources Public.* Fort Collins, 276 p.

Salas, J.D., Yevjevich, V., Lane, W.L., 1980. Applied modeling of hydrologic time series. *Water Resources.* Littleton, 484 p.

6.7   Short Questions

1. Briefly discuss the types of models for the mathematical simulation of karst aquifers.
2. What are black-box models?
3. Indicate the most relevant black-box models that you know.
4. Indicate the convolution integral and briefly explain each term.
5. Discuss the possible impulse responses with daily and monthly data, and their limitations.
6. What is more reasonable as a function of input: rainfall data or estimated recharge? Justify your answer.
7. In deconvolution, why is it advisable to set restrictions on the impulse response?
8. How do you explain that the impulse responses obtained through correlation analysis are different from those obtained with deconvolution?
9. Does it make sense to use deposit models with many elements?
10. What disadvantages are there to pretending to simulate using deposit models with many elements?
11. Give the parameters in an ARMA model and their physical meaning.
12. Explain briefly what an ARIMAX model consists of.
13. Is it justifiable to apply conventional mathematical simulation models to the study of karst aquifers?
14. Are models based on blocks and conduits more decisive than those based on elementary reference volumes?
15. What are the main limitations of simulation models in karst?

6.8   Personal Work

1. Compare the Mero, CREC, BEMER and TRYDEP models and undertake a critical analysis.
2. Apply convolution and deconvolution to series of rainfall and flow in karst aquifers.
3. The processes of AR, MA, ARMA and ARIMA can be applied to the study of karst aquifers. Critically analyse this statement.
4. 'Traditional' models may be applied to the simulation of karst aquifers. What are the advantages and disadvantages?

References

1. Girard, G. (1970). Essai pour un modèle hydropluviométrique conceptuel et son utilisation au Québec. *Cahiers de l'ORSTOM, VII, 1.*
2. Mero, F. (1978). *The MMO8 hydrometeorological simulation system. Basic concepts and operators guide.* Rapport Tahal T/78-02. Tel Aviv.
3. Guilbot, A. (1975). *Modélisation des écoulements d'un aquifère karstique (liaison pluie-débit). Application aux bassins de Saugras et du Lez.* Thèse 3ème cycle, p. 110. U.S.T.L. Montpellier.
4. Bezes, C. (1976). *Contribution à la modélisation des systèmes aquifères karstiques; établissement du modèle BEMER: son application à quatre systèmes karstiques du midi de la France.* Thèse 3ème cycle, p. 135. U.S.T.L. Montpellier.
5. Padilla, A., & Pulido-Bosch, A. (1988). "TRIDEP", un modelo lluvia-caudal aplicado al estudio de acuíferos. *II Congreso Nacional de Geología, II,* 425–428.
6. Padilla, A., & Pulido-Bosch, A. (2008). Simple procedure to simulate karstic aquifers. *Hydrological Processes, 22,* 1876–1884.
7. Salas, J. D., Yevjevich, V., & Lane, W. L. (1980). *Applied modeling of hydrologic time series* (p. 484). Littleton: Water Resources Publication.
8. Yevjevich, V. (1972). *Stochastic processes in hydrology* (p. 276). Fort Collins: Water Resources Publication.
9. Marquardt, D. W. (1963). An algorithm for least squares estimation of non linear parameters. *Journal of the society for Industrial and Applied Mathematics, 2,* 431–441.
10. Marsily, G. de. (1971). Programme de déconvolution DUHAMEL. Notice explicative. *Ecole des Mines de Paris,* p. 28. C.I.G. LHM/N/71/36.
11. Marsily, G. de. (1977). Programme de déconvolution GMDUEX (ex DUHAMEL). *Ecole des Mines de Paris,* p. 18. C.I.G. LHM/RD/77/25.
12. Pulido-Bosch, A. (1988). Estudio de caudales de descarga de acuíferos kársticos mediante deconvolución. *Boletín Geológico y Minero* XCIX-III, 425–431.
13. Cavanilles, A. J. (1795). *Observaciones sobre la Historia Natural, Geografia, Agricultura, población y frutos del Reyno de Valencia* (2ª ed., 1958, 2 T, p. 747). Madrid: CSIC, Inst. Elcano, Zaragoza.
14. Calvo, L. (1908). *Hidrografía subterránea* (p. 289). Catalá y Serra: Gandía.
15. Pulido-Bosch, A., & Benavente, J. (1986). Aplicación de la deconvolución al estudio de la descarga de El Torcal de Antequera. *El Agua en Andalucía, II,* 413–422. Granada.
16. Pulido-Bosch, A., de Marsily, G., & Benavente, J. (1987). Análisis de la descarga del Torcal de Antequera mediante deconvolución. *Hidrogeología, 2,* 17–28.
17. Padilla, A., Pulido-Bosch, A. (1986). El programa "Evadía" para estimación automática de la lluvia útil a nivel diario. *II SIAGA* (Vol. II, pp. 631–636). Granada.
18. Iglesias, A. (1984). Diseño de un modelo para el estudio de descargas de acuíferos. *Modelo Medas. Boletín Geológico y Minero 95,* 52–57.
19. Watson, G. N. (1966). *A treatise on the theory of Bessel functions* (p. 804). Cambridge University Press.
20. Padilla, A., Pulido-Bosch, A., & Mangin, A. (1994). Relative importance of baseflow and quickflow from hydrographs of karst spring. *Ground Water, 32,* 267–277.
21. Padilla, A., Pulido-Bosch, A., & Calvache, M. L. (1995). On the applicability of ARMA models to time series analysis of karstic spring flow. *Comptes rendus de l'Académie des sciences. Série 2. Sciences de la terre et des planètes, 321,* 31–37. (París, 321 (serie II a)).
22. Padilla, A., Pulido-Bosch, A., Calvache, M. L. & Vallejos, A. (1996). The ARMA models applied to the flow of karstic springs. *Water Resources Bulletin, 32*(5), 917–928.
23. Pulido-Bosch, A., & Padilla, A. (1988). Deux exemples de modélisation d'aquifères karstiques espagnols. *Hydrogéologie, 4,* 281–290.
24. Pulido-Bosch, A., & Fernández-Rubio, R. (1979). Los grandes poljes del sureste de la provincia de Valencia. *Acta Geológica Hispánica, 14,* 482–486.

25. Champetier, Y. (1972). *Le Prébétique et l'Ibérique côtiers dans le Sud de la province de Valence et Nord de la province d'Alicante (Espagne)* (Vol. 24, p. 169). Thesis University of Nancy. Sciences de la Terre.

26. Ledoux, E., & Tillié, B. (1985). *Programme NEWSAM. Notice d'utilisation. Aide-mémoire. Ecole des Mines Paris* (p. 20). C.I.G.

27. Pulido-Bosch, A., & Padilla, A. (1987). Evaluation des ressources hydriques de l'aquifere karstique du Torcal de Antequera (Málaga. Espagne). *XXII Congrès A.I.H.* Rome.

28. Kovacs, A., & Sauter, M. (2007). Modelling harst hydrodynamics. In N. Goldscheider, & D. Drew (Eds.), *Methods in karst hydrogeology* (Vol. 1, pp. 201–222). IAH.

29. Bear, J., & Schwarz, J. (1966). *The hydrogeological regime of the turonian-cenomaniam aquifer of central Israel*. Tel Aviv.

30. Bonnet, M., Margat, J., & Thierry, P. (1976). Essai de représentation du comportement hydraulique d'un système karstique par modèle déterministe: application à la Fontaine de Vaucluse. *2ème Coll. Hydrol. Pays Calcaire*, 79–95. Besançon.

31. Besbes, M. (1978). *L'estimation des apports aux nappes souterraines. Un modèle régional d'infiltration efficace*. Paris: Université Pierre et Marie Curie.

32. Ledoux, E. (1980). *Modélisation intégrée des écoulements de surface et des écoulements souterrains sur un bassin hydrologique*. Curie, Paris: Thèse Université Pierre et Marie Curie.

33. Kiraly, L. (1979). Remarques sur la simulation des failles et du réseau karstiques par éléments finis dans les modèles d'écoulement. *Bulletin du Centre d'Hydrogéologie, Neuchâtel, 3,* 155–167.

34. Kiraly, L., & Morel, G. (1976). Remarques sur l'Hydrogramme des Sources karstiques simulé par modèles mathématiques. Bulletin du Centre d'Hydrogéologie. *Université de Neuchâtel, 1,* 37–60.

35. Kiraly, L. (1984). La régularisation de l'Areuse (Jura suisse) simulé par modèle mathématique. In *Hydrogeology of Karstic Terrains. Case Histories. Burger et Dubertret* (pp. 94–99). A.I.H.

Hydrogeochemistry and Water Quality

7

7.1 Glossary

The *main ions* in karst aquifers are: CO_3H^-, $SO_4^=$, Cl^-, NO_3^-, Ca^{2+}, Mg^{2+}, Na^+ and K^+.

Frequency analysis of the conductivity of spring water from karst aquifers can establish the degree of karst development.

The degree of *erosion and removal* of karst terrain can be estimated by the aquifer's hydrological balance and the mineralogic content of its water.

The *ionic ratios* of rMg^{2+}/rCa^{2+} and the rCl^-/rCO_3H^- ions convey a great deal of information about various processes.

Box plots can graphically represent statistics for each sampled point, such as maxima, minima, mean, median and quartiles to help to identify anomalies.

Scatter graphs are a reliable way to compare variables.

A matrix of scatter graphs is of great use in establishing the balance between multiple variables, showing p variables, with plots for each of the p $(p - 1)/2$ possible pairs.

Cluster analysis of the main and factorial components can find any similarities and differences in the data, generating *dendrograms,* giving a good visualization. *Chebotarev's square diagrams* are equally useful.

The deuterium excess **d** (defined as $d = \delta D - 8\delta^{18}O$) and the recharge altitude are related: the higher the value, the greater the value of d. This is known as the pseudo-altitude effect.

Main sources of potential pollution: urban wastewater outlets and run-off.

Urban solid waste, also from pig and poultry farms, urban developments near chasms, sinks and depressions in the karst massif.

A. Pulido-Bosch, *Principles of Karst Hydrogeology*, Springer Textbooks in Earth Sciences, Geography and Environment, https://doi.org/10.1007/978-3-030-55370-8_7

Agricultural practices near poljes, dolines, over-extraction from coastal aquifers, and industrialization.

The necessary protection will be determined by the intensity of land use, the aquifer's recharge regime and its spatial extent, its characteristics and hydraulic properties, its ability to recover, its piezometric level and how it changes, the depth and lateral extent and the possible causes of contamination.

The three fundamentals of land use planning to protect aquifers: the greatest protection is to be afforded to aquifer groundwater/springs; aquifers are categorized by their economic value as a resource; entire area is zoned on the basis of its vulnerability to pollution.

*A **code of good practice*** will list acceptable activities by their potential for contamination for each zone and subdivision, and the recommended controls over both existing and proposed land use.

7.2 Theoretical Considerations

7.2.1 Limestone and Dolomite Waters

The waters associated with carbonate aquifers are of calcium and/or calcium magnesium carbonate types, with the ratio r Mg/r Ca increasing with the dolomite content of the rock, possibly exceeding that of the unit. Classic texts to consult on the matter are: Hem [1], Catalan [2], Custodio [3–5], Appelo and Postma in Chap. 2 [1] and Fagundo [4].

In general, the concentration of salts increases from recharge areas to discharge areas and from surface to depth (in many cases, hydrogeochemical zoning has been proven, with a notable increase in content at depth in areas subjected to overexploitation). Water is rarely saturated in HCO_3^- and in Ca and Mg; the $CO_3^=$ ion content is scarcely detectable.

The major ions are: HCO_3^-, $SO_4^=$, Cl^-, NO_3^-, Ca^{2+}, Mg^{2+}, Na^+ and K^+. The HCO_3^- content of the water is highly variable and depends on several factors that have already been noted. Normally, it ranges between 150 and 350 mg/l, with higher values seen with an increase in $SO_4^=$ and Cl^- of the waters. Its origin lies in the chemical attack of limestone and dolomite. Its concentration increases significantly as a result of the increase in pCO_2, which comes from the decomposition of organic matter in a process of pollution. Ca^{2+} and Mg^{2+} derive from this same attack (although also from gypsum and carnalite), and the rMg/rCa ratio tends to increase from the recharge zones to the discharge zones, with the same lithology, and can indicate marine intrusion (rMg/rCa is close to 5 in seawater; r: values in milliequivalent/l).

The water is subject to the base change (Ca^{2+} and Mg^{2+} by Na^+; icb = [rCl r (Na + K)]/rCl. A decrease supposes a softening, and an increase a hardening. The Ca^{2+} content usually ranges from 50 to 100 mg/l, and it is much more variable for Mg^{2+}, which can range from zero to about 70 mg/l.

The Cl^- ion, if there are no salts, is usually from rainwater, and the values rarely reach 20 mg/l. It is a very stable ion, hence the interest in using it as a guiding element in the methodology of hydrological budgets. When there are Cl salts, it is the dominant ion, being extremely soluble. The relationship $rCl^-/rHCO_3^-$ is usually used to characterize marine intrusion phenomena, since in continental waters it fluctuates around 0.1 (possibly 5) and in marine waters between 20 and 50. It undergoes a temporal evolution as a consequence of the RTE that concentrates it in dry seasons and dissolves it again in the first rainfall after a period of low water.

The $SO_4^=$ ion is not highly abundant in the calco-dolomitic medium, although it is always present. It comes from the dissolution of the gypsum that may be present in small proportions, and from the oxidation of sulphur (pyrite and others, always present in the medium). It is subjected to reduction processes, a fact that can be used as an indicator of organic contamination. The NO_3^- can be present in small quantities due to storms and humic acids. In values higher than 1030 mg/l, it is necessary to establish if there is any contamination by organic matter. This can be reduced to NO_2 and NH_4^+ or vice versa.

As for physical characteristics, the water in a karst medium has a pH close to 7, either slightly acidic or slightly alkaline. This has to be determined at its point of emergence, noting the value in reference to the temperature and stirring the water well. Water temperature is an essential parameter in hydrogeological research. The water in large aquifers has an almost constant temperature throughout the year, corresponding to the annual average of the area (homothermal zone). Hanging water (epikarst aquifer) displays a highly variable temperature over time. It can be a criterion for recognizing whether or not waters in a given system are integrated, with a view to the identifying the various catchments.

Conductivity and dry residue, both referring to a certain temperature, are a very economic and rapid way to establish both the mineralization of the water and its evolution over time. Both are linked by a close relationship (1.5 or 0.7). The important thing about them (between 300 and 700 μmhos/cm and 200,500 mg/l), and especially about conductivity, is that they can be correlated to the content of the predominant ions (HCO_3^-, $SO_4^=$, Ca^{2+}), giving highly reliable values that can be compared to intermediate measurements. This is not the case for minority ions (Cl^-, Mg^{2+}, Na^+ and K^+). The analysis of the frequencies of the conductivity values yields qualitative information about the degree of karstification, as Bakalowicz in Chap. 3 [6] has shown (Fig. 7.1).

The physicochemical characteristics of the waters associated with these aquifers vary considerably over time, given that many factors influence the equilibrium chains, which in turn vary over time, although this variation is within relatively restricted limits. Its spatial variation is equally remarkable.

An aquifer's water is subjected to modifying processes, such as:

- concentration/precipitation
- oxidation–reduction
- ionic exchanges, etc.

The causes of these must be analyzed on a case-by-case basis.

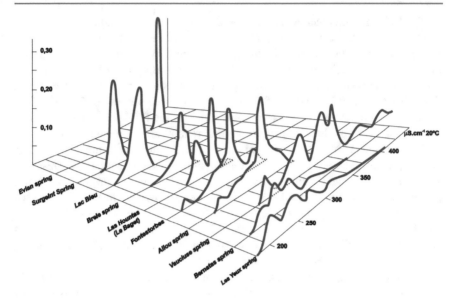

Fig. 7.1 Frequency distribution of the conductivity of various springs. Highly karstified aquifers show a typically multimodal distribution

If we know the mineralization of the waters and the water budget, it is possible to determine the weight and/or the volume mobilized by corrosion. Thus, for example in the Sierra Grossa aquifer, around 40,000 t/year ($\approx$16.000 m^3/year) of CaCO$_3$ are mobilized; that is around 100 kg/km^2/year (40 m^3/km^2/year). Bakalowicz estimates an annual volume of around 600 m^3/year in the Baget karst system of 13.3 km^2 of surface area, referring to total mineralization, which means a volume of 45 m^3/km^2/year of rock evacuated. We can deduce that it takes between 4000 and 20,000 years for an entire, well-organized speleological network to be generated in a karst environment, and this fact is of great interest. Therefore, karst is a dynamic entity, with an elevated permeability K over most of its active life.

Until now, we have considered the aquifer to be pure calco-dolomitic, with no effect from other saline materials. If there are such effects, there will be immediate physicochemical signs of variable spatial extent, dependent on the hydrodynamic circumstances (in the case of Sierra Grossa, Valencia). Nevertheless, in the natural regime any effect by salts may not obvious if, as consequence of the vertical zonality, it is immediately modified by extraction (as in the Mortara in Sierra Grossa).

7.2.2 Data Presentation

The classic graphs by Schoëller-Berkaloff (with its Durov variant), Piper and Stiff are used. The first two have more than one version and have the advantage of quickly identifying possible processes, such as mixes, exchanges, dissolutions and precipitations, and the third just one. Since the physicochemical characteristics of water varies over time, it is necessary to indicate the date of analysis. Unless there is a high density of water points, it is not advisable to make isocone curve plans, given the great heterogeneity of the environment. The most expressive representation is a Stiff diagram for each aquifer point, showing the specific dates. *Star* diagrams have four or more axes extending radially from a central point, and whose values are joined by rectilinear sections on which it is possible to include further statistical parameters of the ion in question. Both star and Stiff diagrams have the disadvantage that when there is great variation in ionic concentration, the scale of each axis has to be changed to avoid points falling beyond the graph's limits.

Maps of isocone graphs involving a mix of waters from several aquifers, and from several zones within the same aquifer, are unacceptable. The analysis of ionic relations rMg^{2+}/rCa^{2+}, $rCl^-/r\,HCO_3^-$, icb. as the most relevant, can be represented in tables or on a plane by means of the punctual value, and this contribution to hydrodynamics is of great interest.

Representations of analytical data, besides maps, are the same as for any type of aquifer, starting with histograms with decimal or logarithmic scales if the amplitude of the concentrations is large. Histograms of several simultaneous samples allow comparisons to be made, although no more than four or five samples can be shown. Other graphs, such as *box plots*, allow the introduction of statistics for each sampled point, such as its maximum, minimum, average, median and quartiles, permitting easy identification of any anomalous values. Comparison of variables has a worthy tool in the form of *scatter diagrams*. The simultaneous relationship between more than two variables can be of great interest in a *scatter plot matrix* [6]. Given the p variables, it consists of generating a scatter plot for each possible $p\,(p-1)/2$ pair of variables. In the lower row, histograms of each variable are represented independently. It is a way to visualize information quickly and to identify similarities and variations.

As for possible data procedures, statistics offers a huge variety that, in general, allow differentiation and discrimination within the great mass of data to identify similarities or differences. The most classic are cluster analyses (of variables and cases or Q and R modes), analyses of the main components and factor analysis. From the resulting representations, processes can be inferred to explain the resulting groupings.

The study of the complexity of the reactions of chemical attack and dissolution of the water system, atmosphere and carbonate rock in the natural environment has advanced notably. A series of codes can quantify the approach in a much safer way than a mere estimate. The most common are WATEQ in its various improved versions, and PHREEQC [7]. They allow the rapid calculation of the saturation indices of calcite, dolomite, gypsum and partial pressures of CO_2 of theoretical equilibrium of groundwater, among many other achievements, as will be seen in the examples that will be discussed.

7.3   Examples of Hydrogeochemical Studies

7.3.1   Box 1: Saline Springs of Sierra de Mustalla (Valencia)

The Mustalla range forms the border between the provinces of Valencia and Alicante (Fig. 7.2) and is made up of locally crushed limestones and dolomites of Cenomenense-Turonense age. It is surrounded by the detritus from the Gandía-Denia plain, apart from at the western end that joins the Almirante mountain range, where it is also made up of carbonate material. From a hydrogeological point of view, this mountain is part of the Sierras de Solana-Almirante-Mustalla unit, covering about 400 km². The main drainage of this unit is along its contact with the detritus of the above-mentioned of Gandía-Denia plain in Chap. 2 [15] and [8].

At about 1, 2 and 3 m asl, respectively, the three most important springs are, from east to west, Salado, Solinar and Aguas (Photos 7.1, 7.2, 7.3 and 7.4). They feed the area occupied by the Natural Park known as the Marsh of

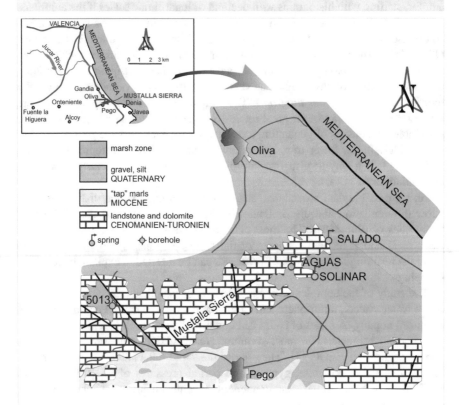

Fig. 7.2 Location of springs and hydrogeological map of the area

Photo 7.1 Sierra de Mustalla and Pego-Oliva marsh, behind rows of orange trees (*Photo* A. Pulido)

Photo 7.2 Downstream of the Aguas spring (*Photo* A. Pulido)

Photo 7.3 The Aguas spring (*Photo* A. Pulido)

Photo 7.4 The river Bullens downstream from the springs of Las Aguas and Solinar, with the Sierra de Segaria in the background (*Photo* A. Pulido)

Pego and Oliva, along with others at the foot of the Sierra de Segaria. They are the source of the rivers Racons, Regalacho and Bullens, whose combined peak flow exceeds several m^3/s. The water of each spring has an elevated temperature, as warm as 29 °C in Salado spring, while Solinar has an intermediate temperature of up to 20.5 °C and the Aguas is at the annual average for the area, 18 °C.

A series of monthly analyses measured the dominant ions in the water of each springs, revealing a wide range of salinity and a notable variation dependent on the sampling period. The most stable ion by far is HCO_3^-. The waters of the Las Aguas spring are of a calcium bicarbonate type and contain less than 1 g/l, while the other two are sodium chloride-dominant. At Salado spring, the highest salt concentrations correspond to the rainy months, at over 17 g/l, while in the other two, the peak is in the dry season.

Of the anions, chlorides dominate. The values lie between 200 mg/l at Aguas and 8500 mg/l at Salado. Sulphates fluctuate widely and are of greater relative importance in the Aguas, peaking in the dry months. The dominant cation is Na, and it can exceed 5000 mg/l in Salado. Mg ranges between 20 mg/l in Aguas up to 450 mg/l in Salado. The Ca ion shows relatively wide fluctuations, and the level is somewhat less in the Aguas, with a maximum there close to 100 mg/l.

At the time of the survey, several hypotheses were proposed regarding the origin of this high salinity. The heights suggested that Salado was the base spring and the two others were overflows, characterized by increasing salinity, flow and temperature from west to east. The considerable flow was explained by the wide extent of the catchment for the aquifer, and the variations at the point of emergence by marked contrasts in transmissivity between the quaternary sediments and karst terrain. Advanced theories about the high salt content include: (1) the dissolution of salts by aquifer water on its transit to the springs. It had been concluded that the only substrate able to contribute so much salt is the Keuper, never before described in this area, and, indeed, subsequent investigation by means of research boreholes has since revealed its presence [9]; (2) a mixing of seawater with the freshwater. This hypothesis is supported by Schoeller's diagrammatic representation of the aquifer, the springs and the sea, which shows certain parallels (Fig. 7.3), as well as the ionic relationships of rMg^{2+}/rCa^{2+} and $rCl^-/rHCO_3^-$ (Table 7.1), pointing the same way. The results of an electric geophysical survey in the sector cannot indicate whether the salt is from current seawater or from water trapped in the geological past in the sediments of the plain, but isotopic studies are able to distinguish the origin. The positive thermal anomaly has its origin in the flow system, greater with proximity to the sea, a fact that would is explained by a mix of the water that emerges.

Cluster analysis has been applied in this area as a means of differentiating that is sensitive to hydrogeochemical description, permitting the eventual identification of processes. Remember that the part of a matrix formed by m

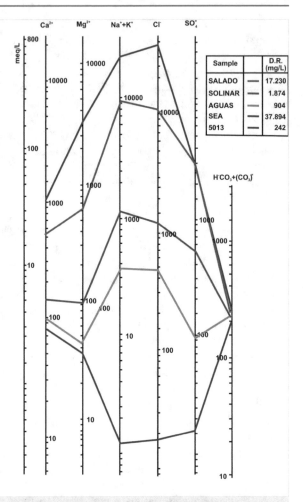

Fig. 7.3 Schoëller-Berkaloff diagram of the freshwater of the aquifer, the sea and the three springs

observations corresponds to an individual, a point in a m-dimensional space [10]. A cluster of cases consists of making n subsets of individuals $p_1, p_2, \ldots, p_n$ in such a way that each individual belongs to one and only one, of these classes from a measure of similarity that varies from case to case, depending on the nature of the problem. A cluster of variables is similarly defined.

In this case, we will use a hierarchical cluster, employing a specific procedure depending on whether it is cases or variables that are involved. For cases, once the distance to use has been fixed, the two nearest points are taken and joined as a group. This is considered a new point whose coordinates are the average of the individual coordinates. The process is repeated with the remaining points—now one fewer—until the whole is in the same group. For variables, the difference is that when two groups are joined to form a larger group, the coordinates of the new point are not the weighted mean but are calculated by one of three possible procedures: (1) minimum distance;

Table 7.1 Values of the ionic ratios and their variation during an observation year

Year	rMg^{2+}/rCa^{2+} sea = 4.3			$rCl^-/rHCO_3^-$ sea = 216		
	Salado	Solinar	Aguas	Salado	Solinar	Aguas
D	0.8	0.9	0.4	17	6.2	2.2
J	1.1	1.4	0.7	15.0	17.2	3.0
F	1.5	1.1	0.6	32.6	15.3	2.3
M	1.6	1.1	0.3	36.7	13.9	2.4
A	1.6	1.0	0.6	45.0	14.0	2.3
M	0.8	1.0	1.0	66.0	12.7	1.4
J	1.6	1.1	0.8	69.4	6.7	2.6
J	1.6	0.9	0.6	58.3	6.9	2.5
A	1.9	0.8	0.7	69.7	6.5	2.7
S	1.3	0.6	0.6	47.0	1.7	2.6
O	1.6	0.7	1.0	48.4	2.4	1.8
N	1.6	1.0	0.5	42.4	11.3	1.9
D	1.3	0.54	0.6	38.9	1.3	2.0
Average	1.4	0.9	0.6	45.8	8.9	2.3

(2) average distance; (3) and maximum distance. In this way, for each level a, a classification of the groups is obtained. The classification evolves from level zero, in which each individual is a group, to all being in the same group. In this example, the correlation has been used for variables, and the distance elucidates the cluster of cases. There are many other possible distances and calculation algorithms available in statistical software packages. The results obtained are usually represented in the form of dendrograms that permit quick visualization.

Figure 7.4 shows a hydrograph of the discharge from the springs of Aguas, Solinar and Salado for more than three years, with Samples 2–24 in chronological order. Samples 1 and 25 were taken before and after the flow recording period, respectively. Peak flow can exceed 7 m^3/s, and even in droughts is rarely less than 1 m^3/s. The analyses were undertaken with 19 samples corresponding to two complete hydrological cycles in order to avoid distortions produced by off-cycle data. A dendrogram of the cluster of variables is given in Fig. 7.5, showing that the greatest similarity corresponds to Cl^-, Na^+ and electrical conductivity, and the smallest to bicarbonates with magnesium, which have a negative correlation. The remaining variables have intermediate values.

Figure 7.6 shows the dendrogram of the variables for the 19 samples extracted over the two complete hydrological cycles. The case dendrogram (Fig. 7.6) identifies two water 'families' at a distance of 2.6 (Cycle 74/75) and 2.8 (Cycle 73/74), although one sample of each cycle (20 and 4) lies outside of this range. In turn, in each cycle's subfamilies are differentiated by affinities that surely have a genetic explanation.

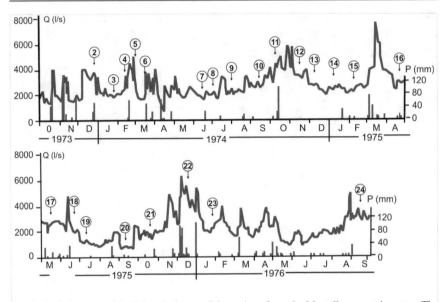

Fig. 7.4 Hydrogram of the joint discharge of the springs from the Mustalla mountain range. The numbers correspond to the water samples analyzed

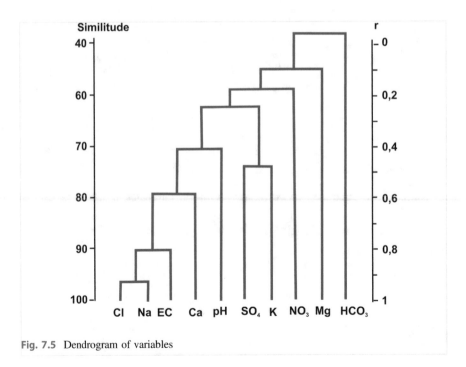

Fig. 7.5 Dendrogram of variables

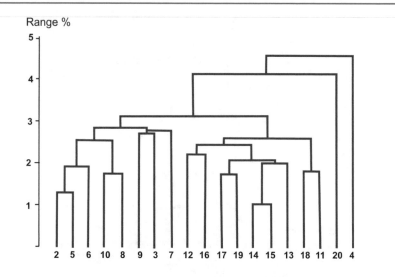

Fig. 7.6 Dendrogram of cases with 19 samples of two hydrological cycles

7.3.2 Box 2: Salado Springs (Sierra Gorda, Granada)

These are the slightly thermal springs of Membrillo (19.4 °C) and Bañuelo (20.4 °C), both perennial, and Fuente Alta, with a greater yet seasonal flow, overflowing after heavy rain. The first two springs are at a altitude of 545 m and are 150 m apart, while Fuente Alta is at 568 m. The overall average flow rate is around 85 l/s. These springs emerge on the eastern edge of Sierra Gorda, which in this sector forms a syncline whose centre is comprised of up to 500 m of Cretaceous marls (Fig. 7.7), bounded by the Depression of Granada's post-orogenic material. This is a sector affected by numerous fractures oriented northwest-southeast.

The main discharge of the Sierra Gorda aquifer is along its northern edge at an altitude close to 500 m in Chap. 1 [28] and [11]. Although the outcrops are apparently distant from Sierra Gorda and part of a minor outcrop of Jurassic limestones surrounded by Neogene and quaternary material, including calcarenites and Tortonian conglomerates, the prime contributing area is indeed the Sierra Gorda, which explains the water's thermomineral characteristics (Fig. 7.8). While the most widespread hydrogeochemical facies is calcium bicarbonate, in the Sierra Gorda, it is magnesium calcium, in Bañuelo sodium calcium bicarbonate-chloride and in Membrillo calcium–magnesium bicarbonate-sulphate.

These aspects are clear from the results of principal component analysis (PCA) of the analytical samples extracted from water points across the Sierra Gorda (Fig. 7.9). Both the elevated salt content (with electrical conductivity values approaching 1000 microS/cm in Bañuelo and 850 microS/cm in

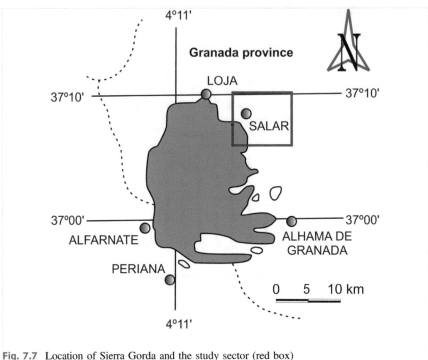

Fig. 7.7 Location of Sierra Gorda and the study sector (red box)

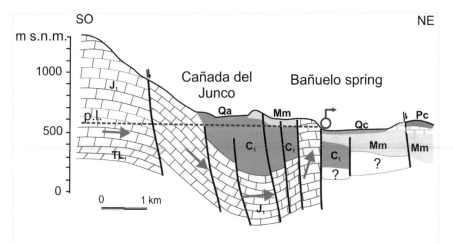

Fig. 7.8 Hydrogeological cross section explaining the springs of the Salar. Unit of Sierra Gorda: TL1, Liassic dolomite; J1, Jurassic limestone; C1, Cretaceous marl and marl-limestone. Granada Depression: Mm, silt and marl from the Upper Miocene; Messiniense lacustrine limestone; Pc, detritic materials from the Plio-Pleistocene; Qc, colluvial Quaternary; Qa, alluvial Quaternary

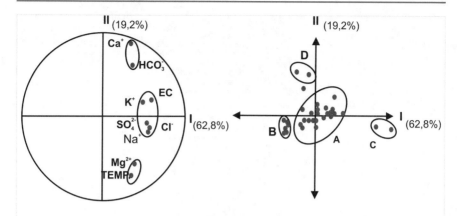

Fig. 7.9 Spatial distribution of variables (left) and cases (right) of the water data of the 33 points of the Sierra Gorda aquifer. **a** Springs in the Loja-Riofrío sector; **b** Springs in the southern sector; **c** Springs in the Salar sector; **d** Springs in the epikarst sector

Table 7.2 Isotopic concentration of Bañuelo water, and the range of variation in other springs in the Loja sector (June 1993)	Isotopic parameters	Bañuelo	Other Loja springs
	$\delta^{13}C$ (‰) [PDB]	−5.39	−82.32 a −9.65
	$\Delta^{14}C$ (‰)	−762 ± 4	−350 a −460
	pMC (%)	23.6	54 a 64.1
	^{3}H (TU)	<1	6.7 a 8.7
	$\delta^{18}O$ (‰) [SMOW]	−7.71	−7.99 a −8.43
	^{222}Rn (Bq/l)	19.8	4.3 a 10.6
	^{226}Ra (Bq/l)	0.0013	0.0076 a 0.0123

Membrillo) and the highly variable chemistry are explained by deeply circulating water and dissolution of salts, very possibly of the Triassic substrate. The isotopic data also point to the uniqueness of Bañuelo (Table 7.2), at least —apart from $\delta^{18}O$ which, although less negative, suggests a feeding area in common to all the springs of this area. The tritium content indicates that it is older water, not current seawater.

Monitoring the parameters over an entire hydrological year made it possible to compile Table 7.3, highlighting the narrow variation in electrical conductivity, indicative of a fairly homogeneous and well-mixed medium. However, in the measured data, the distribution curve is bimodal, interpreted at the time as due to drainage of a low-mineralized surface fraction (linked to Miocene calcarenites) into the subterranean water of Sierra Gorda.

Table 7.3 Statistics of the physical and chemical variables of Bañuelo, 1986–87

	Q (l/s)	Temp (°C)	pH	Cond (µS/cm)	Cl^- (mg/l)	SO_4^{2-} (mg/l)	HCO_3^- (mg/l)	Na (mg/l)	Mg^+ (mg/l)	Ca^{2+} (mg/l)	K^+ (mg/l)
Sample n	11	6	5	51	51	51	51	51	51	51	51
Average	44	20.2	7.25	982	134	148	227	75	33	77	3.1
Minimum	33	19.4	7.10	930	119	127	203	61	27	68	2.8
Maximum	58	20.7	7.41	1018	153	168	205	83	45	88	3.4
Variance, typical	9.15	0.46	0.12	23.19	8.25	12.84	16.02	7.18	3.40	4.72	0.17
Coef. var (%)	20. 7	2.3	1.7	2.4	6.2	8.7	7.1	9.6	10.2	6.1	5.6

7.3.3 Box 3: Sierra de Gádor, Pilot Area

7.3.3.1 Hydrogeochemical Characterization

The southern edge of the Gádor mountain range, which covers 320 km^2, (Fig. 7.10) and its extension into Campo de Dalías have been the subject of numerous hydrogeological and hydrogeochemical studies for more than thirty years. Situated in the internal zone of the Betic Cordillera, the area is comprised of the Alpujárride formation of the Gádor unit and, above all, the Felix unit, integrated by a metapelite at the bottom and a thick carbonate series with intercalations of calco-schist (Photos 7.5, 7.6, 7.7, 7.8, 7.9, 7.10, 7.11 and 7.12) In the Gádor unit, the somewhat marbled limestone and dolomite can exceed a depth of 1000 m, while the Felix unit is seldom over 100 m deep. There are also Miocene and Pliocene calcarenites and extensive alluvial fans. Mechanical surveys show that under the Pliocene calcarenite are sandy marls that merge into marls, and these can reach a depth of 700 m.

The Piper diagram shown in Fig. 7.11 was constructed for the data for January 1993 and a seawater sample. The resulted show a great diversity of facies in the study area, ranging from calcium bicarbonate to sodium chloride, with evidence of alignments that clearly indicate a mixing of dissimilar characteristics; the sea is clearly one extreme, but there is also enrichment of

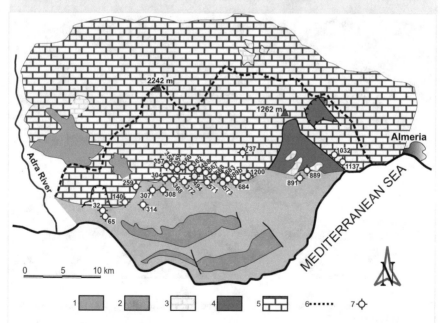

Fig. 7.10 Hydrogeological scheme of the southern edge of the Sierra de Gádor and Campo de Dalías. 1: quaternary materials; 2: Pliocene calcarenite; 3: Miocene calcarenite; 4: Felix carbonates; 5; Gádor carbonates; 6: Hydrological divider; 7: Sampled and its number

Photo 7.5 Reddish phyllites of the Felix unit (*Photo* A. Pulido)

Photo 7.6 View of the central part of the Sierra de Gádor with alluvial fans in the foreground (*Photo* A. Pulido)

Photo 7.7 View of the carbonates of the Gádor unit with accumulation of tailings, testimony of the intense mining exploitation that took place since the end of the nineteenth and twentieth centuries; Vícar Rambla (*Photo* A. Pulido)

Photo 7.8 One of the numerous hydraulic check dams in this case, which mark the southern edge of the Sierra de Gádor (*Photo* A. Pulido)

Photo 7.9 One of the boreholes of sector III of Campo de Dalías, producing more than 100 l/s (*Photo* A. Pulido)

Photo 7.10 Panoramic view of the Sierra de Gádor, with wind turbines at the summit (*Photo* A. Pulido)

Photo 7.11 Spectacular fold
in the carbonates and tracings
of the western half of the
southern edge of Gádor
(*Photo* A. Pulido)

sulphates of a different origin. There has clearly been marine intrusion and
extrusion, modifying both processes and mixtures (Fig. 7.11).

To identify the processes better, refer to Chevotarev's quadrangular diagrams in Fig. 7.12, representing the percentage of bicarbonates against the
sum of the percentages of chloride, sulphate and nitrate; that is, the 'sweet'
versus the 'saline' fraction. Deviations from the theoretical freshwater–saltwater mix are due to modifying processes ing the cationic content. This
results in vertical displacement in the samples. It is evident that the samples
are indeed displaced, indicating an increase in Ca + Mg and a decrease in
Na + K; that is, inverse ion exchange.

Given the probable presence of sulphates that are not of marine origin, a
diagram to represent the sum of the percentages of HCO_3 and SO_4 can be
constructed to represent 'fresh' water. Fresher water is aligned to the theoretical freshwater–seawater mixing line; those with an intermediate position
with respect to their anionic content indicate a loss of Ca +Mg and a gain of
Na + K, interpreted as a direct exchange. Also evident is an increase in the

Photo 7.12 Recovering a borehole in the El Ejido area (*Photo* A. Pulido)

percentage of Ca + Mg, accompanied by a loss in the percentages of Na + K in the saline water, equivalent to an inverse exchange. This process ceases when the exchange capacity is exhausted in a stabilized mixture. The spatial distribution of the Mg content (meq/l) is depicted in Fig. 7.12.

7.3.3.2 Isotopic Data

Studying the isotopic composition of groundwater provides a great deal of hydrogeological data of interest, such as the tritium in the Sierra de Gádor aquifers. Data were obtained from a sample extracted in September 1993, whose values are shown in Fig. 7.13; Table 7.4. The analyses were carried out at the Environmental Studies Laboratory of the Hungarian Academy of Sciences (Debrecen, Hungary). A particle counter β was used to perform the analysis, and its detection limit was is 1 UT.

The tritium content gives information on the presence of water that has recently infiltrated into the aquifer. However, before discussing tritium values in this context, any input of tritium from rainwater should be taken into account. Since the tritium in water comes mainly from thermonuclear explosions that have occurred after 1952, in order to interpret tritium concentration values, it is essential to establish its presence in the area's rainfall over the last forty years; that is, since nuclear tests have been carried out in the atmosphere. There are no direct measures of tritium in the rainfall across the study area. However, on a global scale and in various areas of the Iberian

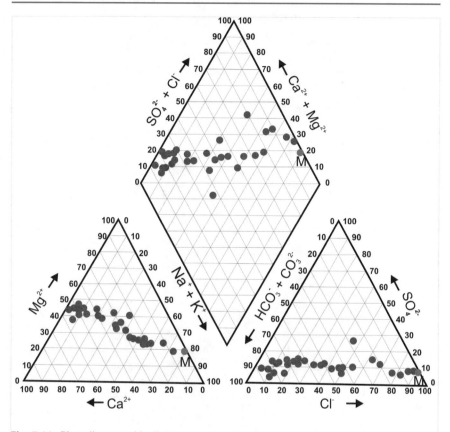

Fig. 7.11 Piper diagram with all water samples. M: seawater

Peninsula, the distribution of tritium in rainfall displays only a narrow variation in concentration, enabling the reconstruction of a distribution curve of tritium in rainfall at any point.

The data values for Sierra de Gádor were obtained by annual correlation until 1963, with a series of average values in Central Europe in Chap. 2 [17] and, until 1991, from the Thonon data of the International Atomic Energy Agency of Vienna. The data in Table 7.4 indicate that the annual concentration of tritium in the area under study has been over 10 UT for the last forty years, after reaching a maximum of 1324 UT in 1963. Samples containing no tritium indicate the absence of water from rain that has fallen in that time. The concentration of tritium found in the waters of the carbonated aquifers of Campo de Dalías are very low. Many of the samples show less than 1 UT; thus, they are pre-1953 waters or, alternatively, are old waters with a very small proportion of recent water.

If the ^{14}C content (% of modern carbon) is represented against the tritium content of the various samples analyzed, two trends are evident (represented

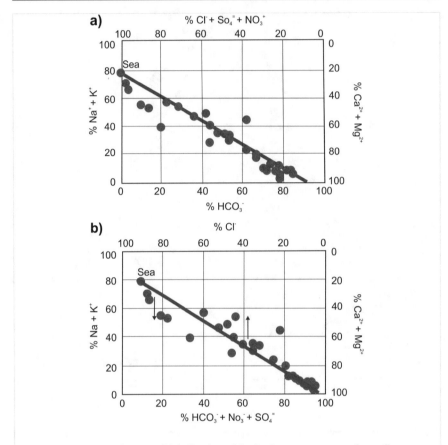

Fig. 7.12 Chevotarev diagram with indication of the freshwater–seawater mixture line

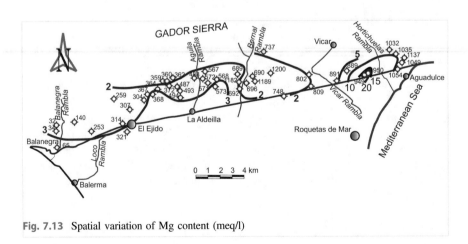

Fig. 7.13 Spatial variation of Mg content (meq/l)

Table 7.4 Annual concentrations of tritium (in UT) in rainwater

Year	Europe	Thonon	Gádor	Vera
1991		15	10	
1990		17	11	
1989		29	16	
1988		25	14	
1987		24	14	
1986		22	13	
1985		22	13	
1984		22	13	
1983		24	14	
1982		29	16	
1981		39	20	
1980		33	18	
1979		59	29	
1978		86	40	
1977		74	35	
1976		117	53	
1975		171	76	
1974		113	52	66
1973		126	57	39
1972		145	65	47
1971		228	101	100
1970		189	84	73
1969		206	91	83
1968		217	96	102
1967		213	94	132
1966		247	109	118
1965		437	190	221
1964		1637	700	611
1963		3106	1324	1284
1962	700		301	350
1961	110		50	48
1960	145		65	65
1959	450		195	220
1958	300		131	140
1957	125		57	55
1956	100		46	43
1955	35		19	15
1954	300		131	140
1953	25		14	10
1952	20		12	8

by two lines), which tend to define the limits of the results in pre- and post-pumping eras, intersecting around 2 UT and 40% of modern carbon (Fig. 7.14). There is no value higher than 8 UT (Table 7.5), implying that the extracted waters have a reduced contribution from the precipitations in recent years. The pluviometric recharge in this environment arrives ready mixed, with both pre-existing waters and the reserve water in the zones of the aquifer

Fig. 7.14  ^{14}C—Tritium ratio

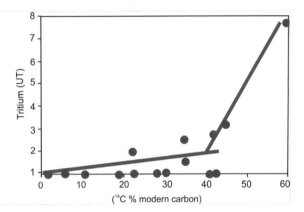

(^{14}C % modern carbon)

Table 7.5 Percentages of recent and old water in the mixed water

Sample	(pmC)	^{3}H (UT)	^{14}C (pmC)	X	100 − X
950	34.4	2.5	32.8	2.8	97.3
304	22.3	<1.0	21.5	1.1	98.9
1054	44.5	3.2	42.8	3.5	96.5
1032	41.6	2.8	40.0	3.1	96.9
314	10.2	>1.0	9.3	1.1	98.9
150	21.9	2.0	20.3	2.2	97.8
259	42.1	1.0	41.6	1.1	98.9
372	19.2	>1.0	18.4	1.1	98.9
748	5.6	<1.0	4.7	1.1	98.9
32	34.8	1.5	33.9	1.7	98.4
1200	59.7	7.7	56.8	8.5	91.5
586	1.4	<1.0	0.4	1.1	98.9
839	30.3	<1.0	29.6	1.1	98.9
362	41.5	<1.0	41.0	1.1	98.9
567	27.5	<1.0	265.8	1.1	98.9
321	17.9	1.0	17.1	1.1	98.9

^{14}C: ^{14}C concentration in the oldest fraction of water
X: percentage of the most recent fraction
100 − X: percentage of the oldest fraction

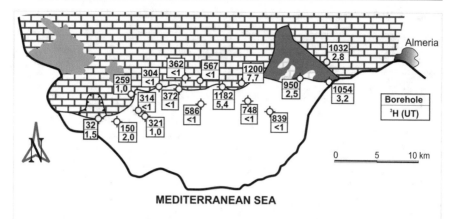

Fig. 7.15 Tritium concentration (UT) of the borehole water samples in the study area

that are exploited. This mixing frequently reduces the tritium concentration and masks the presence of modern recharge.

As can be seen in Fig. 7.15, the easternmost sector of this areas displays higher values, indicating a greater influence of modern recharge (Boreholes 1182, 1200 and 950); thus, there is a preferential circulation of groundwater corresponding to the current marine intrusion detected in Boreholes 1032 and 1054. There are several discrepancies between the ages calculated from ^{14}C and ^{3}H [12]. Some measures of tritium content indicate recharge since 1953, while the value of ^{14}C indicates an age of around one or two thousand years (e.g. Boreholes 950 and 1032). This result can be explained by a mixing of waters of dissimilar age.

Mazor et al.'s model for calculating the percentage mix of recent and old water [13] can be applied, based on: (1) average post-1953 recharge water containing about 90 UT, along with 90% modern carbon; thus, 1 UT is accompanied by approximately 1 cfm; and (2) the ^{14}C measured in mixed groundwater corresponds to the sum of the ^{14}C of the most recent fraction and that of the oldest fraction. The percentage of the oldest fraction in the mix can be deduced from the ^{14}C concentration of the oldest. Analysis of the data obtained at the edge of Sierra de Gádor shows that, in all cases, the percentage of old and recent components in the mix exceeds 90% (Table 25).

As has been demonstrated, as well as establishing the age of water, recharge areas and differentiating the flow systems, among others [14, 15], isotopes can provide the criteria by which to distinguish between alternative hydrogeological processes.

The analytical results (Table 7.6) have an accuracy of ± 0.05 per thousand for ^{18}O and ± 1 per thousand for ^{2}H. The reference standard used was V-SMOW. The samples were taken monthly between October 1991 and March 1993 in four surveys at the foot of the mountains. The sampling network (Table 7.7) was expanded twice for ^{18}O, with an accuracy of ± 0.2

Table 7.6 Results of isotopic analyses (in ‰) of groundwater samples

Date	140		372		567		737		1137	
	^{18}O	D	^{18}O	D	^{18}O	D	^{18}O	D	^{18}O	D
10/1991	−8.18	−52.4	−8.14	−50.0	−7.97	−50.5	−8.93	−56.5	−8.52	−54.5
11/1991	−8.17	−52.6	−8.36	−52.0	−8.23	−51.9	−8.78	−56.8	−8.35	−54.0
12/1991	−7.96	−55.8	−8.13	−54.8	−7.77	−54.2	−8.85	−58.5	−8.23	−52.6
01/1992	−8.11	−52.0	−8.36	−52.8	−8.14	−50.5	−8.63	−55.6	−8.30	−53.7
02/1992	−8.09	−54.4	−8.37	−53.1	−8.32	−50.9	−8.74	−56.4	−8.40	−56.1
03/1992	−8.07	−47.9	−8.39	−52.9	−8.32	−52.2	−8.68	−58.3	−8.45	−55.3
04/1992	−8.03	−50.4	−8.69	−53.3	−8.27	−51.6	−9.03	−56.6	−8.96	−54.3
05/1992	−8.23	−53.6	−8.06	−51.8	−7.92	−53.5	−8.73	−55.0	−8.31	−56.0
06/1992	−8.03	−53.0	–	–	−8.11	−50.4	−8.80	−56.4	−8.56	−52.8
07/1992	−8.00	−53.5	−8.00	−51.2	−8.00	−50.0	−8.30	−56.1	−8.20	−53.4
08/1992	−8.00	−52.4	−8.00	−51.9	−8.00	−51.8	−8.40	−54.5	−8.30	−53.0
09/1992	−8.10	−51.0	−8.10	−52.2	−8.00	−52.4	−8.60	−52.2	−8.20	−53.0
10/1992	−8.10	−50.3	–	–	−7.90	−52.2	−8.60	−56.8	−8.30	−56.6
11/1992	−8.00	−51.9	−8.20	−49.8	−7.90	−54.3	−8.50	−55.8	−8.50	−52.3
12/1992	−7.80	−52.8	−8.00	−49.4	−8.00	−53.7	−8.40	−55.4	−8.30	−53.2
01/1993	−8.30	−49.8	−8.20	−52.0	−8.00	−51.8	−8.60	−54.2	−8.50	−52.5
02/1993	−8.00	−51.4	−8.30	−50.5	−8.00	−49.5	−8.80	−54.4	−8.20	−53.3
03/1993	–	–	−8.30	−51.2	–	–	−8.80	−55.1	−8.30	−53.5

per thousand. The ^{18}O of the rainwater (Table 7.8) was also sampled in three rainwater stations in the mountains, located at various heights; some data from previous studies were also used [16].

In relation to the D content of the rainwater, various features detected are shown in diagrammatic form in Fig. 7.16 relative to that in Fig. 7.15. The discrepancies are interpreted as due to weather events in the sampling year and their influence on the local topography [17].

There is a correlation between the excess values of deuterium d (defined as $d = \delta D - 8\delta^{18}O$) and the extent of its presence, with a correlation coefficient under $r = 0.46$: the higher the value, the greater the value of d, as a general trend (Fig. 7.19). This is what is called the pseudo-altitude effect [18, 19]. The average isotopic gradient is calculated directly from the dimensions of the samples taken on both sides. The altitude effect has been used to estimate the average level of recharge in the aquifer at the southern edge. The results obtained using the O^{18} altitude relationship for both lines and for the whole area are shown in Figs. 7.18 and 7.19. It follows that rainfall of Mediterranean origin would contribute more to recharge than those of Atlantic origin.

Regarding the isotopic composition of the groundwater, the representation of δD versus $\delta^{18}O$ for the same period as the previous sample is shown in Fig. 7.20. The tendency (dotted line) is to converge towards the MMWL for $\delta D = 65$ per thousand, which seems to suggest a meteoric origin for the

Table 7.7 ^{18}O Content in groundwater samples (in ‰)

Sample	^{18}O	
	June 1993	September 1993
32	−7.80	–
140	−8.62	–
144	−9.02	−9.40
150	−8.40	–
259	−8.35	–
304	−8.47	−8.56
307	−8.23	–
314	−8.67	−8.73
321	−8.60	–
362	−8.16	–
372	−8.37	−8.63
567	−8.23	–
571	−8.65	–
683	−9.25	−9.45
684	–	−9.43
748	−7.35	–
891	−8.01	–
950	−7.84	–
1032	−8.73	−8.63
1054	−8.83	–
1137	–	−8.95

Table 7.8 Content in ^{18}O, excess deuterium d (in ‰) of rainwater samples. Data 1–10 are taken from Araguás (1991)

Sample	Altitude (m)	^{18}O	D	d
1	910	−8.23	−45.4	20.4
2	1065	−6.29	−30.6	19.7
3	1540	−7.15	−40.4	16.8
4	1845	−11.30	−72.7	17.7
5	1260	−6.64	−36.6	16.5
6	550	−0.68	7.9	13.3
7	1790	9.34	−57.0	17.7
8	2030	−8.61	−51.6	17.3
9	1950	−10.90	−72.40	14.8
10	30	−4.25	−22.7	11.3
11	1438	−8.16	–	–
12	710	−6.70	–	–
13	610	−6.05	–	–

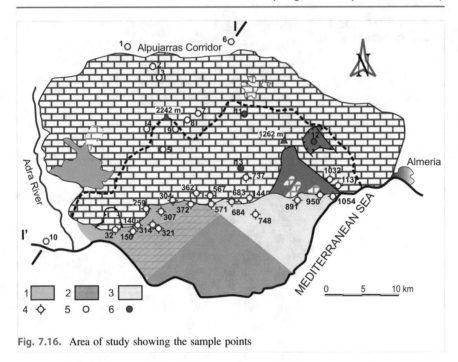

Fig. 7.16. Area of study showing the sample points

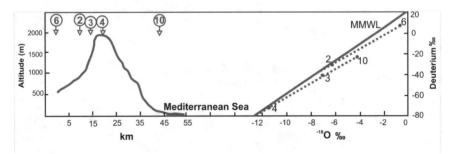

Fig. 7.17 Values of the isotopic composition of the samples along the I-I'section in Fig. 7.16. The circle numbers correspond to the station code of the indicated figure. MMWL is the Mediterranean Meteoric Water Line

water. The line obtained by linear regression with $r = 0.6$ has a 4.5 slope. The wide dispersion is attributed to evaporation during run-off, and in the soil itself, before infiltration (Fig. 7.21).

Figures 7.22 and 7.23 indicate values of $\delta^{18}O$ along the contact of the carbonates of the Sierra with the Plio-Quaternary materials seen the field in June and September of 1993, with concentrations ranging from −7.35 to −9.45 per thousand. The detail of these values provides much information on recharge mechanisms and preferential flows when we compare the values to

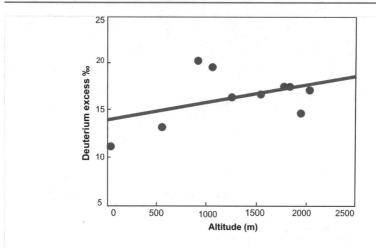

Fig. 7.18 Excess values of D in the rain samples as a function of the sampling altitude

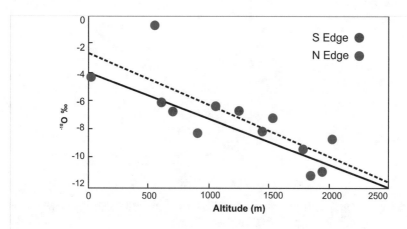

Fig. 7.19 ^{18}O in the rainfall against altitude on the north and south slopes

those in other climatic conditions. For example, the waters of Boreholes 144, 683 and 684, with significantly lighter isotopic contents, reaching −9.45 per thousand in ^{18}O in some cases, are indicative of recharge in a fast-flow and high-altitude area. It is low-salt water, with an TSD between 330 and 400 mg/l. Tritium data (7.7 and 5.4 UT in Boreholes 144 and 683, respectively) indicate that the water is younger than its surroundings.

According to the linear relationship obtained from the $\delta^{18}O$—dimension in the rainfall data of the southern highlands (Fig. 7.24a), taking into account the isotopic composition of the groundwater, the greatest aquifer recharge takes place between 1100 and 1700 m, which is of notable interest when applied with a view to constructing systems to allow increasing infiltration

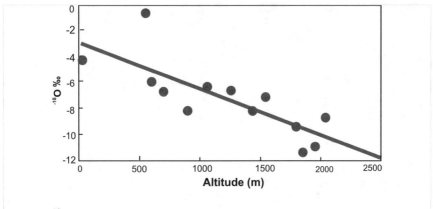

Fig. 7.20 ^{18}O versus altitude for all rain samples from the sierra

Fig. 7.21 Relationship between D/18O for the rainfall on the sierra

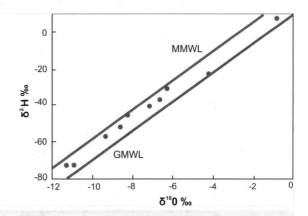

from surface run-off. If the entire massif is considered, the altitudinal strip is substantially similar (1200–1800 m; Fig. 7.24b).

Isotopic data can also provide information on the possible presence of a marine intrusion. The conservative behaviour of D and ^{18}O in water molecules, together with that of Cl in groundwater, allows us to identify the origin of salinity effectively. Figure 7.25 shows the Cl ^{18}O ratio in the borehole water. Where marine intrusion is the cause of the increase in salinity, the analyzed sample lies on the theoretical freshwater–seawater mixture line. This is the result with samples 950, 1054 and 1137, located in the easternmost sector, with the highest salinity. Bu contrast, when the increase in salinity is due to washing out evaporitic deposits, there is no change in the stable isotopes of the infiltrating waters. It should be noted that in Fig. 7.25 there is no relationship detected between the ^{18}O and Cl content to that of other samples.

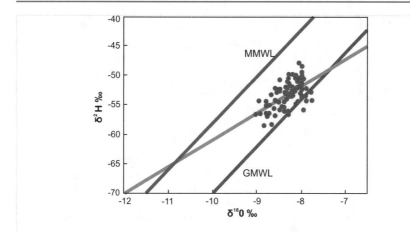

Fig. 7.22 Relationship between D/ ^{18}O for the waters of the boreholes of the mountains during the period October 1991—March 1993. GMWL: Global Meteoric Water Line. MMWL: Mediterranean Meteoric Water Line

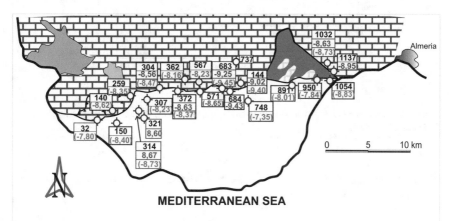

Fig. 7.23 Special distribution of the ^{18}O content of the waters of some boreholes whose number is in bold. The data to June, in parentheses, and September 1993, in a rectangle

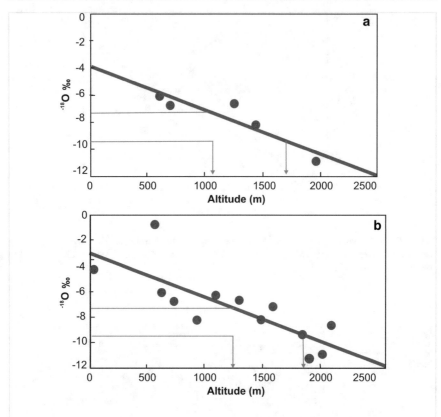

Fig. 7.24 Main recharge areas according to isotopic groundwater data. **a**: southern edge; **b**: Sierra de Gádor as a whole

Fig. 7.25 Cl versus ^{18}O in the groundwater samples of the aquifer

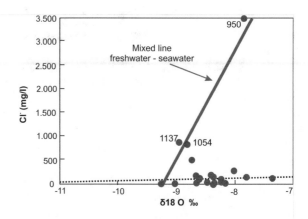

7.3.4   Box 4: The Crevillente Aquifer

7.3.4.1   Geological Framework

This aquifer has been intensively exploited for many years, with an annual extraction exceeding 20 hm^3, while the estimated annual recharge is between 6 and 10 hm^3. As a result, the piezometric level has decreased by up to 40 m/year and the water's salt content has increased significantly. This water has been used to irrigate 9000 ha of fairly profitable table grapes for many years. The increase in operating costs and the deterioration in the water quality mean that its profitability has been questionable since the start of the present millennium and even before.

The aquifer of the Crevillente mountain range occupies 87 km^2 of surface and is on the Sub-Betic-Prebetic contact (Fig. 7.26). The oldest outcropping materials [20, 21] are clay and gypsum Triassic Keuper facies. Jurassic dolomite and limestone occupy most of the area's sierras (Crevillente, Ofra, Argallet, Rollo, Pelada and Reclot), and there are lateral changes in the facies, interruptions in the deposits, sinsedimentary fractures and palaeokarstification [22].

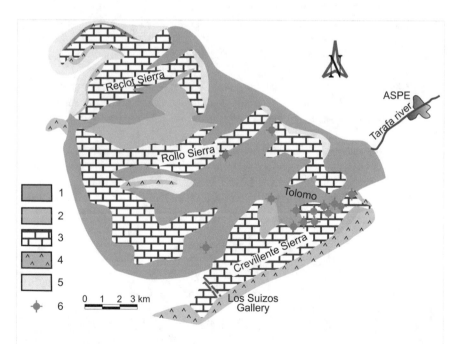

Fig. 7.26 Aquifer of the Sierra de Crevillente and its surroundings: (1) Quaternary sediments; (2) Cretaceous marl; (3) Jurassic carbonate; (4) Trias; (5) Prebetic; (6) Borehole and number

Cretaceous marl occupies the synclines between the anticlines of Jurassic limestone and dolomite. Eocene materials occur only on the southern edge of the Sierra de Crevillente, represented by green marl and limestone with flint, about 25 m thick. Miocene materials emerge very locally (Abanilla-Monte-Alto), and quaternary sediments occupy large areas, with glacis and alluvial between La Romana and Aspe. The underlying ridged Prebetic materials are constituted of limestone, calcarenite, marl and marly limestone of Eocene and Miocene age [23].

7.3.4.2 Hydrogeology

The aquifer materials are Liassic fissured and karstified limestone. The impermeable substratum and the northern and southern edges is of Triassic clay, with gypsum and Prebetic marl. Specific flow rates from the area's boreholes range from one to 100 l/s/m. Transmissivity values vary between 3000 (Sierra del Rollo) and 50,000 m^2/day. The average effective porosity is estimated at 2.2%. The recharge from rain infiltration is estimated at 610 hm^3 in an average year and 16 hm^3 in a wet year. Pumping extracted 16 hm^3/year in the southwest, 10 hm^3/year in the Tolomó, in the southeast of the area and just one hm^3/year in the south of the Sierras del Rollo and the Cava, and the same in the area between the Galería de los Suizos and Tolomó.

The Galería de los Suizos is 2316 m long, and its first 700 m are through marl and clay and the rest through Jurassic carbonates. Its rising flow can reach 1 m^3 s^{-1}, but it soon dries up. Subsequently, 12 boreholes were drilled inside, with flows of 45–75 l s^{-1} m^{-1} and transmissivity values of around 7500 m^2/day. The average annual drop in the level within the gallery has been 8 m/year (Figs. 7.27 and 7.28). In the Tolomó sector in the period 1962–94, the piezometric level decreased by 290 m.

7.3.4.3 Hydrogeochemical Evolution

Due to the intensive exploitation, there has been a continuous increase in the salinity of the pumped water, which has changed from calcium bicarbonates to sodium chlorides in some sectors of the calcium sulphated facies. The explanation for this saline increase is the possible presence of vertical hydrogeochemical zoning, with maximum salinity at depth from the evaporites of the Triassic substrate. Exploitation has not affected the aquifer uniformly, as a whole, and zoning has appeared that is dependent on both the exploitation and the nature of the substratum.

Cl is mainly responsible for the increase in salinity (Fig. 7.29), although sulphate is also significant. However, the relationship of the latter to conductivity (Fig. 7.29b) shows a notable dispersion that can be attributed to the lower homogeneity of the distribution of sulphates in the sector (Fig. 7.30).

Figure 7.31 shows the changes in the saturation index with respect to calcite (SIc), dolomite (SId) and gypsum (SIg), calculated with the WATEQB code [24]. Only two of the samples are clearly oversaturated with calcite and

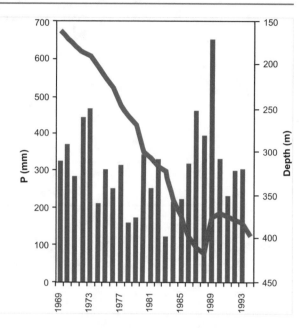

Fig. 7.27 Piezometric record of a representative borehole of the Tolomó corresponding to January of each year, with superimposed rainfall values

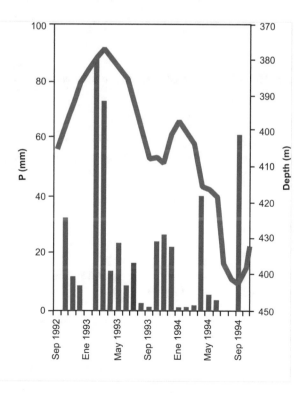

Fig. 7.28 Detail of the monthly piezometric record from September 1992 to September 1994

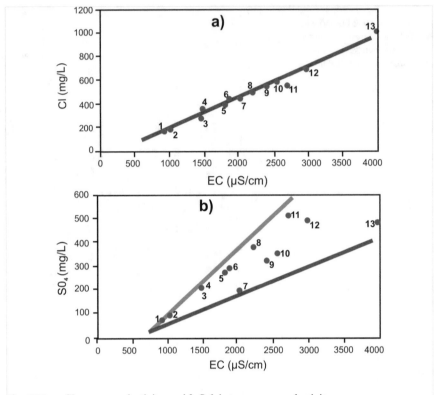

Fig. 7.29 **a** Cl versus conductivity; and **b** Sulphates versus conductivity

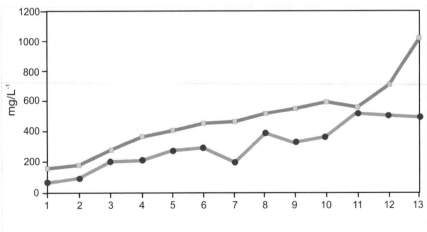

Fig. 7.30 Graph of the Cl and SO_4 content of the water at each survey

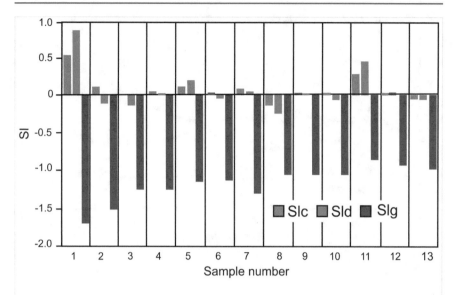

Fig. 7.31 Bar graph showing saturation index values for calcite (SIc), dolomite (SId) and gypsum (SIg)

dolomite: one in the sector of the Rollo, corresponding to water of low salinity; and 11 in the Galería de los Suizos, relatively enriched in sulphates and calcium. All samples are undersaturated with respect to gypsum.

There is a remarkable correspondence between the sulphate, magnesium and strontium ions (Fig. 7.32), due to their common origin in the underlying evaporite sediments, although Sr may be related to the celestine ($SrSO_4$). Mg can come from the dissolution of the dolomite from the aquifer matrix by sulphate-rich waters, which would lead to calcite precipitation and would soon reach oversaturation (a common-ion effect) and parallel enrichment in Mg, according to classic models. Sr has a similar origin and precipitation–dissolution processes, suggested by many authors [25]. Sr enrichment can be related to the time that the water spends in the aquifer, thus is used as an environmental tracer.

The dissolution of the carbonates of the aquifer enriches the waters in Ca, Mg and Sr. The precipitation of calcite in the successive stages generates a relative increase of Sr. These times of residence can also be interpreted as quantities of relative flow, so that slow flows increase the duration of the contact with the rock, as happens in the deeper sections of the Crevillente aquifer. If we take as a reference the waters of lower concentration in Ca, Mg, and Sr with these values, for each sample, we can calculate the percentage of variation of Sr with respect to the sum of Ca + Mg. The increase in these variations would be explained as a greater maturity, equivalent to a longer duration and, consequently, greater interaction with the rock.

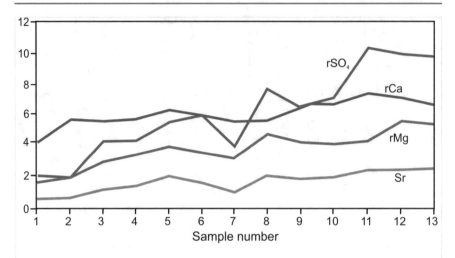

Fig. 7.32 Graph of SO_4, Ca, Mg, and Sr (in meq l-1) content in each survey

Figure 7.32 shows the evolution of the content in Ca, Mg and Sr ions. The last two have similar behaviour while Ca has certain deviations, probably due to saturation in calcite and subsequent precipitation. Figure 7.33 shows the evolution of MI in relation to the maturity index, calculated with the expression:

$$MI = \frac{\Delta Sr}{\Delta Ca + Mg}$$

The results are shown in Table 7.9, taking Sample 1 as the reference. The most relevant aspect of Fig. 7.33 is that the MI does not follow a coherent evolution with the increase in salinity, indicative of other processes altering the proposed simple model. In fact, three types of behaviour can be identified: Points 1 and 2, located in the Sierra del Rollo, show low maturity, characteristic of recently infiltrated waters of low salinity; Points 6 and 7, in the central sector and in the southern strip of the aquifer, show a relatively low concentration of Sr and sulphates; and finally, following the proposed model, the rest of the waters.

According to the above, the hydrogeochemistry of this aquifer is the result of synergy of at least four factors: (1) hydrodynamic control, related to pumping and continuous piezometric descent; (2) lithological control, presence or absence of dolomite in the aquifer; (3) proximity of evaporites, controlled by tectonics; and (4) climate control, i.e. amount of infiltrated water, closely related to the amount of precipitation. Their combined influence on the hydrogeochemistry of the aquifer causes the continuous degradation of water quality. The structural factor is decisive in the double sense of

Table 7.9 Calculation of the maturity index (MI)

Sample	rCa	rMg	Mr	rCa + rMg	ΔCa=	ΔSr	MI
1	4.14	1.97	0.5	6.11			
2	5.64	1.97	0.6	7.61	1.50	0.1	0.066
3	5.59	2.92	1.1	8.51	2.40	0.6	0.250
4	5.64	3.29	1.4	8.93	2.82	0.9	0.319
5	6.29	3.87	2.0	10.16	4.05	1.5	0.370
6	6.04	3.54	1.6	9.58	3.47	1.1	0.316
7	5.59	3.21	1	8.80	2.69	0.5	0.185
8	5.94	4.77	2.1	10.71	4.60	1.6	0.348
9	6.64	4.20	1.9	10.84	4.73	1.4	0.296
10	6.79	4.20	2	10.81	4.70	1.5	0.321
11	7.54	4.28	2.5	11.82	5.71	2	0.350
12	7.19	5.68	2.6	10.99	4.88	2	0.409
13	6.69	5.43	2.6	12.12	6.01	2.1	0.349

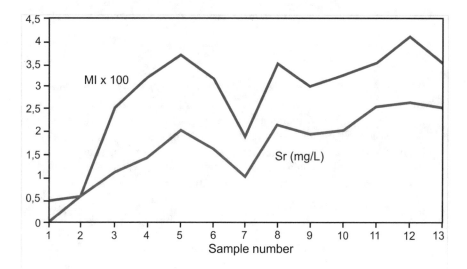

Fig. 7.33 MI and Sr graph in the waters of each survey

intensifying the water–rock interaction by prolonging the duration and promoting the washing out of chloride ions and sulphates from the evaporite substrate, which also constitutes the lateral barrier of the system.

From a hydrogeochemical point of view, the presence of an aquifer component is deduced: (1) at the time of this study, the northern sector (area of the Sierra del Rollo) is less intensely exploited and has water of good quality and close to recharge; (2) the eastern sector of the Tolomó shows a rapid fall in level, with evaporite materials nearby and progressive salinization of the water; (3) the central sector of the Tolomó shows little evaporite influence; and (4) the western sector, Galería de los Suizos, has similar characteristics to the eastern sector but with high concentrations of sulphates.

By way of final discussion, the conceptual model of this karst aquifer is widely represented in the Betic mountain ranges, typically affected by tectonics to a considerable extent, with a great thickness of carbonates and the presence of evaporite materials at depth, which gives a fairly homogeneously karstified set with hardly any variation of the hydrodynamic parameters at depth (an 'homogeneously karstified' aquifer that can be regarded as acting like an equivalent porous medium). CO_2 of deep origin, together with a long history of emersion and palaeokarstification, may explain this apparent homogeneity. It is important to highlight the fact that the complex geological structure of the area promotes the disconnection of sectors. Descending water finds the impermeable substrate closest to the surface, giving rise to sub-aquifers, leading to notable piezometric jumps and water with a very variable saline content.

7.3.5 Box 5: Sorbas Gypsum

The temperature of the main spring water (n 1) varies between the measuring points, ranging from 19.8 to 22 °C in Chap. 2 [17], which points to a positive thermal anomaly. The temperature of the water in the survey of the Venta was 21 °C. By contrast, the temperature of the waters of the small epikarst springs is more variable and is influenced by the outside temperature (Las Viñicas, No. 2, at 14.5 °C and Cueva del Tesoro, No. 3, at 16.5 °C, in October).

The conductivity of the water is also variable, ranging between about 5000 microS/cm in the main spring, 2400 microS/cm in surveys for commercial sales of gypsum and about 3000 microS/cm in the vineyards. The Stiff diagrams are included in Fig. 7.34. These are calcium sulphated facies water, and the analytical results from these aquifers are included in Table 7.10.

The waters of the Mills are those with the highest salt content. The differentiating ions are Cl and Na, which at the time of the study were interpreted as identifiers of lateral feeds from the beach sands (Cantera formation), as it seemed to indicate a concentration of 25 meq/l in water captured in Tertiary loamy sands.

Estimates drawn from the average saline concentration of the main source and its average flow (100 l/s) are that chemical denudation is responsible for 9000 tonnes per year of gypsum, equivalent to 260 m^3/km^2, much higher than the average denudation estimated for carbonate aquifers.

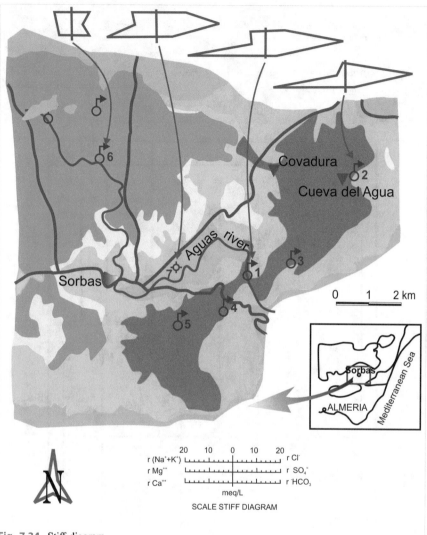

Fig. 7.34 Stiff diagram

Table 7.10 Analytical results of Sorbas gypsum: hydrogeochemical data (mg/l)

Point	Date	Ca	Mg	Na	Ka	Cl	SO₄	HCO₃	NO₃
1	10/1960	32.8	7.2	6.7	0.1	7.8	34.3	3.9	–
	08/1962	33.0	6.0	6.7	0.1	8.2	33.2	4.2	0.0
	04/1965	31.0	7.3	8.7	–	9.1	33.5	4.1	0.0
2	10/1960	30.2	0.8	0.7	0.0	1.0	27.0	1.7	–
	08/1962	30.0	3.0	1.1	0.0	1.2	28.9	2.2	0.12
	04/1965	29.6	1.75	1.17	–	1.3	26.8	4.5	0.0
Water siphon	09/1982	27.0	5.0	0.9	0.1	1.0	27.3	2.6	0.0
3 (Tesoro)	10/1980	30.0	2.8	1.0	0.1	1.7	27.0	2.5	–
Gypsum salt	04/1980	30.1	4.5	2.21	–	2.1	31.0	2.96	0.24

7.3.6 Box 6: Hydrogeochemistry of the Vallada Area (Valencia)

In the Schoëller-Berkaloff diagram (Fig. 7.35), the main ionic contents of the two water samples analyzed of the six collected in Vallada are synthesized. The first corresponds to the furthest point accessible upstream of the entrance

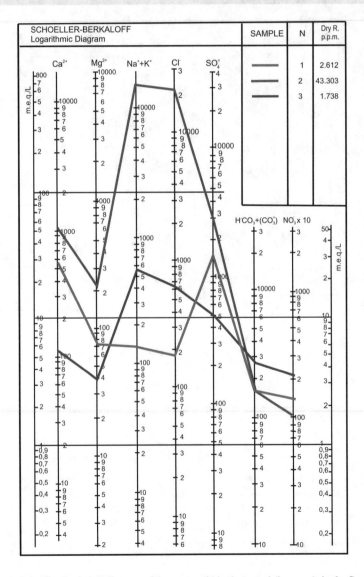

Fig. 7.35 Schoëller-Berkaloff diagram of the water within the tunnel (in green), in the Saraella (in red) and in the well in Fig. 40 in Chap. 2 (blue)

to the tunnel, with sulphated sodium facies; and the second to the Saraella spring, with more than 43 g/l of dry residue and chlorinated sodium facies. The remaining four samples were taken at separate points within the tunnel and gave very similar conductivity values (about 2800 microS/cm). There is no variation in HCO_3 content, although the concentrations of Mg and SO_4 increase. The considerable increase in saline content is interpreted as due to halite dissolution of the Keuper at the furthest accessible point of the tunnel, and to the mixing of brines from the dissolution of halite that is also Triassic, yet within the massif of Sierra Grossa. These would increase the area of the basin's slope to the Saraella.

In this sense, water in the boreholes in the northern front of Sierra Grossa near the depression of Canals presents an anomaly in terms of its content of ions such as Cl, Na and SO_4. The water at one of these points was also analyzed and the results placed on the aforementioned diagram, confirming that they also present sodium chloride facies.

7.3.7 Box 7: Salt Evaporites of Fuente Camacho

7.3.7.1 The Data

Fuente Camacho is a variety of diapiroid of Triassic evaporites (Keuper facies) containing a series of superficial karst forms, including flooded dolinas —'lagoons'—and a brine spring that supplies salt, giving its name to the small neighbouring village (Figs. 7.36 and 7.37, and Photos 7.13, 7.14, 7.15 and 7.16).

The hydrogeochemistry of the Fuente Camacho upwelling was studied through a series of samples taken from its most representative aquifers (Table 7.11) followed by hydrogeochemical modelling, to determine the processes and identify possible flow systems and potential pathways in this environment, from infiltration to the point of emergence. The stages were:

1. Vadose water or water from the most superficial strip, with minimal water–rock contact. The well dug in a dolina near Fuente Camacho (No. 9) was considered representative, like the Pilas de Fuente Camacho (No. 10), which drains an outcrop of ophite with gypsum. Also included is all water not in contact with hypersoluble salts.
2. Deeper circulating water that reaches the saturated strip and that has had water–rock contact time. This is the main drainage of the system. The two lagoons, Grande (No. 7, without hypersoluble salts in its environment) and Chica (No. 8, with hypersoluble salts) are included.
3. Water with hypersoluble salts. This includes the salinas themselves, with three representative samples:

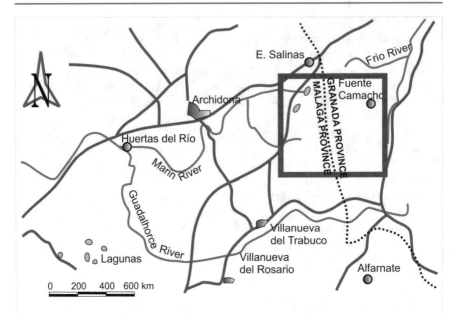

Fig. 7.36 Location of Fuente Camacho and its surroundings

(a) Vadose or near-surface water (Salinas II, Point 11)
(b) Point 12, from the brine spring
(c) Point 13, the saline spring after a drought (Salinas Ib).

In addition, to compare to other sectors of the Antequera Triassic, samples were taken from the Lomas sector (Nos. 1 and 2) and the Loma del Yesar (No. 3, Agua cave).

7.3.7.2 Data Procedures

For the data procedures, the first consideration with water of such a high salt content is that the ways in which salts reach the water are both ionic and molecular. This is what is known as coupled ions, as opposed to free ions. The content of one or the other is determined by the respective equilibrium constants at a given temperature. The ions coupled in natural waters are $CaHCO_3^-$, $CaCO_3^\circ$, $CaOH^+$, $MgHCO_3^+$, $MgSO_4^\circ$ and $MgOH^+$, at the margin of the associations of Na and K. The ions coupled in natural waters are $CaHCO_3^-$, $CaCO_3^\circ$, $CaOH^+$, $MgHCO_3^+$, $MgSO_4^\circ$ and $MgOH^+$, at the margin of the associations of Na and K.

The sequence of equations derived from these balances must be based on the activity of each species, related to molality [x] by means of the activity coefficient:

$$<x> = \gamma x[x]$$

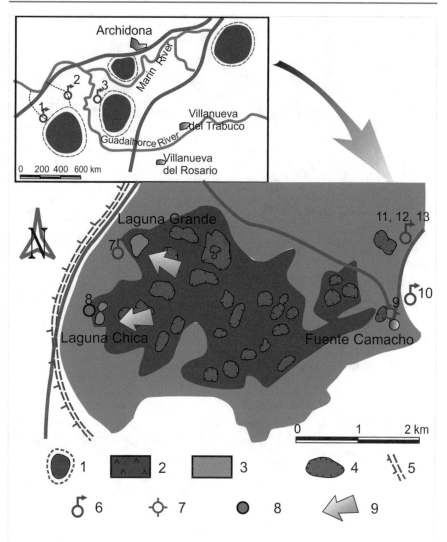

Fig. 7.37 Study area of Archidona and detail of Fuente Camacho: (1) Diapiroid; (2) Essentially Triassic gypsum; (3) Triassic clay, sandstone and ophite; (4) Doline; (5) Paleopolje limit; (6) Spring; (7) Well or borehole; (8) Surface water; (9) Main drainage of aquifer

The activity coefficient is very close to that in very dilute solutions, and it decreases if the ionic strength increases. This is why the difference between activity and molality can be high in solutions with a high ionic content. With an ionic force of 0.05, typical of sulphated solutions in gypsum karst, the activity of the SO_4 ion reaches 50% of its molality. The ionic activity is calculated by an iterative process, since the activity coefficient depends on the ionic force (I) of the solution and, in turn, on the concentration of ions in the

Photo 7.13 Fuente Camacho salt flats, March 2014 (*Photo* A. Pulido)

Photo 7.14 Detail of the
tapestry of halite crystals in
Fuente Camacho (*Photo* A.
Pulido)

Photo 7.15 One of the numerous gypsum quarries, subsequently filled with inert material (*Photo* A. Pulido)

Photo 7.16 Laguna Grande de Antequera, formed on the Triassic gypsum (*Photo* A. Pulido)

Table 7.11 Physical and chemical parameters of aquifer water analyzed (January and February 1989) [28]

Sample	Sector	Q l/s	Tra C	pH	Eh mV	EC	HCO3	SO4	Cl	Na	K	Mg	Ca
1 Warrior	Las Lomas	<1	13.4	7.22	−10	2540	186	1910	65	92	4.7	79	628
2 Angostura	Las Lomas	5	16.3	6.75	3	2450	232	1790	63	68	6.5	81	588
3 Water Cave	Yesar	35	16.4	7.35	−16	2400	278	1749	61	77	7.1	140	528
7 Laguna Grande	Fuente Camacho	25	16.3	6.90	8	2750	272	1810	124	116	4.7	121	564
8 Laguna Chica	Fuente Camacho	?	9.1	8.29	−68	8650	120	2340	1296	850	23.6	490	915
9 Dolina Camacho	Fuente Camacho	<1	10.9	7.60	−30	1550	221	805	97	78	14.4	63	300
10 The Batteries	Fuente Camacho	3	14.5	7.02	2	1300	289	734	29	53	5.3	38	286
11 Salt mines 11	Fuente Camacho	<1	8.3	7.56	−27	26,750	343	935	5914	3830	35.4	76	476
12 Salt mines 1a	Fuente Camacho	2–3	12.6	6.44	32	168,000	362	2119	42,623	28,200	110.2	148	777
13 Salt mines 1b	Fuente Camacho	2–3	16.0	6.00		190,000	282	4040	164,542	100,412	157	473	1094

solution. To solve the equations of equilibrium in this theory, we use the PARQUIMIC program to obtain results similar to WATEQ, WATSPEC, GEOCHEM, but with a simplification of the phases present in the medium.

PARQUIMIC solves the equations by the continuous fraction method, under the initial assumption that $[Ca^{2+}]=[Ca^{2+}$ total], and so on for all ions. Calculations of the activity coefficients are made using the Debye-Huckel equation for ionic forces <0.1 and the Davis correction for higher values of I. The equations for the various equilibrium constants have been collected or elaborated by Plummer [26], Langmuir [27] and Dreybrodt in Chap. 2 [2], among others. In addition, PARQUIMIC incorporates restrictions that, without altering the calculations, accelerate the resolution. Among others, ionic coupling between Cl, Na, K can be avoided to limit the phases to the most relevant, from a hydrogeological point of view (calcite, gypsum, anhydrite, dolomite, halite), to ignore the calculations of the ionic forces of OH^-, H^+, $CaOH^-$, and so on, yet activating $CaSO_4^0$ in karst waters, the effect of the common ion and the exchanges in ionic form.

The data obtained are essentially intended to obtain the values of the saturation indices of the phases selected according to the hydrogeological environment. These indices are expressed in logarithmic form:

$$SATXY.... = \log(<x><y>..../Kxy),$$

where Kxy is positive for oversaturated water and negative for undersaturated water. Table 7.12 shows the initial laboratory data in which dissolved salts are expressed as their total content, without taking ionic coupling into account. The saturation indices shown in Table 24 have been calculated according to the activity of the free ions present in the solution. For water with an ionic strength greater than 0.5—which is the case of Salinas—the results are only approximate.

7.3.7.3   Hydrogeochemical Approach to Gypsum Karst

To be adequate, hydrogeochemical modelling of an aquifer must establish the representative hydrogeochemical structure of the aquifer and correctly identify the lithology–chemical composition interrelation. Calculation of the saturation indices for the most representative minerals of the medium is the first step. In this specific case, we start from a model of soluble blocks and layers and carbonates through which infiltrating water passes on its way to the spring, as the carbonates include essentially limestones and dolomites. The most soluble levels are halite (less abundant) and gypsum. The study of the variations of the saturation indices SATCAL, SATDOL, SATGYP and SATHAL in the water samples provides information on the water's state of equilibrium as it passes through the aquifer.

A first simplification is that the blocks are only of calcite and alabaster. The water circulating in such an aquifer has the equilibrium conditions of the

Table 7.12 Values of saturation indices for the different species in the samples studied

Sample	Sector		Fion	pPCO$_2$	SATCAL	SATGYP	SATHAL	SATDOL	SATHAL
(1) Warrior	Las Lomas	1	0.057	2.10	−0.28	0.24	−0.18	0.09	−6.78
(2) Angostura	Las Lomas	2	0.050	2.51	−0.7	−0.10	−1.88	0.05	−6.85
(3) Water Cave	Yesar	3	0.055	2.04	−0.35	0.52	0.75	−0.04	−6.90
(7) Laguna Grande	Fuente Camacho	7	0.058	1.60	−0.32	0.09	−0.22	0.00	−6.41
(8) Laguna Chica	Fuente Camacho	8	0.141	3.48	−0.29	1.07	2.00	0.12	−4.54
(9) Dolina Camacho	Fuente Camacho	9	0.033	2.40	−0.79	0.48	0.46	−0.40	−6.64
(10) The Batteries	Fuente Camacho	10	0.027	1.67	−0.78	0.07	−0.50	−0.44	−7.32
(11) Salt mines 11	Fuente Camacho	11	0.254	2.26	−1.10	−0.56	0.46	−0.69	−3.18
(12) Salt flats 1a	Fuente Camacho	12	0.985	1.23	−0.90	−0.45	−1.37	−0.56	−2.37
(13) Salinas 1b	Fuente Camacho	13	6.314	1.47	−0.63	−0.63	−1.10	−0.05	−0.04

CO_2–$CaCO_3$–$CaSO_42H_2O$–H_2O system. If the water passes through a carbonate level, this would be detected by the increase in Ca^{2+} from the calcite, but there would be no variation if it passed through gypsum. On the other hand, if the water flowed through alabaster, the opposite would be true, as far as the Ca is concerned. For practical purposes, the first case would be like introducing a certain gypsum content into the system:

$$[CaSO_4] = [SO_4^{2-}] + [CaSO_4^0] = [Ca^{2+} \, gypsum] \, (1)$$

Figures 7.38 and 7.39 show the variation found in the saturation index of gypsum, compared to SATCAL and SATHAL, allowing us to track the hydrogeochemical path between the groups of samples already defined. To visualize the precipitation [29] and dissolution conditions, the program BALANCE [30] is used, which solves the mass exchange equations.

If we take Direction 1 in Fig. 7.39, which identifies the route from the infiltration zone to an evacuation zone, we observe the constancy in the calcite saturation index and a notable increase in Ca from gypsum until it reaches saturation. The incorporation of Ca from carbonates is minimal, but the increase is assumed by the gypsum. This increase in gypsum dissolution is accompanied by calcite precipitation, as shown by the BALANCE program (Table 7.13). This loss of Ca from calcite is not well reflected in Fig. 7.39. Since the amount of Ca from carbonate rocks remains approximately constant, Ca ions must come from the dissolution of dolomitic rocks. We face the incongruent precipitation of calcite with dolomite dissolution. This is the process of dedolomitization of a karst aquifer in gypsum.

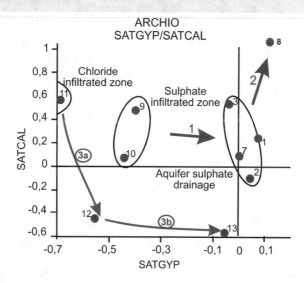

Fig. 7.38 SATGYP-SATCAL relationship

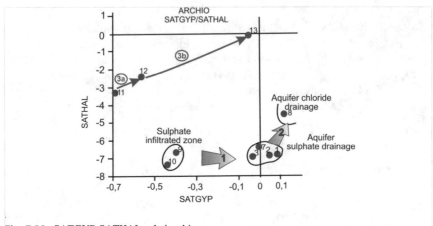

Fig. 7.39 SATGYP-SATHAL relationship

Table 7.13 Results of the BALANCE program: **a** mmol/l, apart from 11, 12, 13 × 10; **b** values at Δmmol/l

A						
Sample	C	S	Cl	Na	Mg	Ca
Sulphated infiltration strip (9,10)	4.186	8.016	1.781	2.837	2.078	7.307
Sulphated drainage from the aquifer (1, 2, 3, 7)	3.965	18.904	2.207	3.833	4.328	14.389
Chlorinated drainage of the aquifer (8)	1.961	24.375	36.507	36.957	20.165	22.818
Chlorinated infiltration strip (11)	0.562	0.974	16.659	16.652	0.313	1.187
Aguacero salinas (12)	0.593	2.207	120.06	122.609	0.609	1.938
Salt flats Estiaje (13)	0.462	4.208	463.50	463.574	1.947	2.728
B						
	1	2	3A	3B		
Calcite	−5.8	−13.5	5.0	−172.3		
Dolomite	2.3	15.8	13.4	13.4		
Halite	0.4	34.3	1034.0	3434.4		
Alabaster	10.9	5.5	12.3	20.0		
CO_2 gas	1.1	−20.2	−10.5	140.5		
Ca/Na exchange	0.3	−0.6	12.7	−14.7		

Direction 2 corresponds to a calcium sulphated water passing through levels of hypersoluble salts (Fig. 7.40). The SO_4^{2-} content increases, as well as the value of the saturation index with respect to the gypsum, allowing greater dissolution of the gypsum. But the main increase occurs in the presence of Ca of carbonate origin (Fig. 7.40). The calcite is the first to

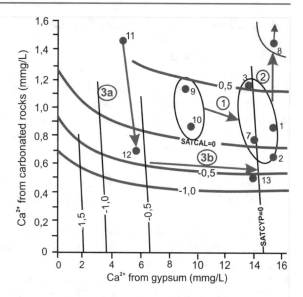

Fig. 7.40 Sequence related to Ca of carbonate and gypsum origin

precipitate, as the water has had to be aggressive with respect to the alabaster and dolomite in its course (Table 7.13b).

Direction 3 is characteristic of an area where hypersoluble salts dominate in the composition of groundwater. Point 11 is representative of the most superficial layer and Point 12 is the spring of the Salinas after a shower of rain. The contact between water and rocks, in this case, is very brief. In this case, there is a sudden drop in Ca from carbonates, but with dilution of gypsum. It is a situation of high saturation (Fig. 7.40). The high ionic strength of these solutions allows high contents of Ca and SO_4 to be maintained in solution with water that is still aggressive towards these minerals. Dissolution of calcite, dolomite, alabaster and halite results.

From Points 12–13, representative of the passage from conditions of very short duration water–rock contact to others of longer contact—Direction 3b—supposes an enrichment in Ca from the gypsum, reaching conditions of saturation in a linear way (Fig. 7.40) until the point where gypsum and halite precipitate. There is massive precipitation of calcite, visible in the existing travertines, although the water is still aggressive towards the other minerals. Gypsum and halite are also precipitated locally, as this does not depend on calcite precipitation.

7.4 Pollution and Potential Sources

7.4.1 Background

Upwelling karst water is considered to be highly vulnerable to rapid and widespread contamination by pollution by various substances. Its self-cleaning power is very low, in general. This aspect has been addressed by various studies; most recently, the final report of the COST project, Action 65, presents a modern and updated synthesis in Chap. 3 [11] of the main pollutants:

- *Surface discharge* is often responsible for the contamination of the water supply itself. Discharge into rivers that feed karst aquifers can cause disaster (as at Ontinyent).
- *Landfill* is usually carried out with no precautions, and in no case is it *controlled*. The leachate is highly polluting, and decomposition in anaerobic medium generates CH4, hence possible spontaneous combustion.
- *Pigs and poultry farms*: These are often situated on karst terrain, as this is not suitable for much other agriculture. They are highly active sources of pollution, although with a limited radius. There have been many examples of such pollution throughout Spain since the 1970s.
- *Urbanizations:* Normally without sewage networks, these settlements have begun to populate karst massifs. The cesspits (elemental septic tanks) are a problematic source of organic, biological and detergent pollution, etc.; too often, the water supply is sited next to a cesspit.
- *The chasms, sinks and depressions of karst massifs* are where dead livestock are traditionally thrown, and at times of epidemic, this can infect large numbers of beasts.
- Agricultural activity on poljes, dolinas, intercalated marls, and so on, constitutes a source of *extensive* contamination that is common in these massifs.
- Heavy extraction of aquifer waters in coastal zones risks the ingress of seawater, and there have already been numerous examples across Spain.
- All types of industry.

The best defence against pollution is preventive one, since any regeneration, at best, is extremely expensive—and it is often unfeasible. That is why the most effective and important work that can be carried out is *dissemination of information and awareness-raising activities* about the serious danger of any polluting activities. The media are could best placed to contribute an environment of citizen responsibility in the face of this problem, which has become endemic due to technological progress. Once it has been created, there are strategies to fight pollution in Chap. 3 [11], which we will not go into deeply here, among other reasons because each case merits a specific study.

In the case of hydrocarbon contamination, in the event of an accident, the procedure to be used is to collect up the oil with shovels and burn it. This is applicable to the karst environment when there is a developed soil, and the time factor is key. If there is no soil, the only way is to construct a pumping barrier to prevent the contaminant's spread into the aquifer, and simultaneous burning. In any case, the cost may exceed €10 million.

With regard to protection, rather than to protect the water source itself, which is complex, it is more effective to protect the area around the source; although not clearly protected under Spanish legislation (but not the case in many other countries), parameters for each water supply's protection can be established on the basis of a series of guidelines:

- the intensity, regime and location of the recharge
- the hydraulic characteristics and properties of the terrain
- its self-cleansing ability
- its piezometric level and changes over time
- its depth and lateral extent
- the possible causes of contamination.

There are three levels of protection:

- in need of immediate protection (land to acquire) within a radius of 100 m
- in close proximity (1000 m for hydrocarbons in fractured terrain)
- in need of distant protection (the whole catchment area).

In a fractured substrate, it is recommended to site a water supply a minimum distance of 30 m from a cesspit to avoid biological hazard. In any case, water chlorination is an effective fight against biological contamination, although certain viruses are known to be resistant.

7.4.2 Hazards in Karst Groundwater

7.4.2.1 Main Impacts

Any human activity influences the environment. Appropriate measures must be taken to reduce the impact of these activities to the minimum, hence the need to study carefully their possible effects. The following will synthesize part of what was elaborated in the final report of the COST-65 Action in relation to this topic in Chap. 3 [11]. International or global risks to karst waters are presented by the long-term effects of polluted transboundary air, due to industry, traffic, domestic heating, agriculture and so on; the consequences of climate change; and radioactive accidents, such as Chernobyl. The difficulty with these potential risks is protection through local or regional regulation; only measures involving international cooperation can mitigate the impact.

At a local or regional level, a general differentiation of activities can be into: infrastructure; industrial activities; and agriculture. These do not have to contaminate the aquifers if they are reasonably designed, but they pose a *hazard*:

- if exceeding a certain threshold
- in the event of accident
- the cumulative effect of low-toxicity substances
- in very vulnerable environments of a karst system (sinks and sectors).
- highly karstified terrain.

The hazard can affect both the quality and the quality of aquifer water. Table 7.14 summarizes the main dangers to karst waters as a consequence of human activities.

Table 7.14 Impact of human activities on groundwater

Activity	Process	Likely negative impact
Infrastructure development		
Domestic wastewater	Migration of pollutants to groundwater	Microorganisms, NH4, NO3, organic tracers
Production	No wastewater filtration addressed	No filtration, self-purification
Storage		
Transport	Septic tank filtration	
Disposal	Sewer filtration	
Transport systems	Infiltration of mineral oil and other contaminants	Salt, hydrocarbons, heavy metals
Highways and highways	Discharges from paved areas	Pesticides, microorganisms, NH4
Railways		Miscellaneous pollutants
Accidents	Filtration of contaminated water	
	Spillage and filtration	
Construction	Destruction of the protective cover	Increased vulnerability
In general	Artificial drainage	Reduction in quality and quantity
Tunnels	Leachate and migration	Organic, metals, NH4, SO4, Cl,
Reservoirs	from landfills and other	pesticides, hydrocarbons
Municipal purification	discharges	Pesticides, hydrocarbons
elimination	Leachate and migration	NO and SOx, organic, micropollutants,
Removal of slurry	Fall of contaminating	heavy metals, indirect impacts: acid rain
from sewage treatment plants	aerosols	and forest damage.
Storage and use of household chemicals		
Emission of air pollutants (traffic, heating)		
Industrial activities		
Mining	Coating reduction	Increased vulnerability, metals
Exploitation	Lowering the water level	Acidification, radioactivity,
Treatment and	Leachate from ores,	hydrocarbons
Concentration	tailings and	Hydrocarbons, heavy metals, salt
Dumps	Grinding residue	Radioactivity
Industrial plants	Leachate and infiltration	Hydrocarbons, heavy metals, salt. Salts,
Construction	of dangerous substances	hydrocarbons, various pollutants
Operation	Leachate and migration	

(continued)

Table 7.14 (continued)

Activity	Process	Likely negative impact
Demolition ***Nuclear activities*** Production of nuclear elements Nuclear power plants Radioactive waste Disposal of solid waste Disposal of liquid waste Emission of airborne pollutants	Leachate and infiltration Subfloor injection In shallow aquifers In rivers Pollutant aerosols	Heavy metals, chlorinated hydrocarbons, indirect impacts (acid rain, e.g.)
Agricultural and forestry activities		
Manure and slurry Collection Storage Application Forage effluents Pesticide application Storage of fertilizers and pesticides Irrigation Intensive use of the medium Deforestation	Leachate and infiltration Application and infiltration Leachate Leachate Leachate and slurry Leachate Leachate and infiltration Erosion	Microorganisms, NH4, NO3, K Fe, Mn, organic acids, BOD. Pesticides NO3, pesticides Salinization, NO3 NO3, pesticides
Forestry Use of mechanical equipment Application of insecticides Emission of airborne pollutants	Leachate and filtrations of lubricating oils and gasoline For agricultural practices	Ammonium, pesticides, methane

Infrastructure development The main problems are associated with waste in liquid or solid form, with vehicle, traffic and construction work in general. The poor absorption and retention of karst materials mean that the polluted water generated by traffic, collectors and treatment plants has easy access to the saturated strip. Construction works themselves can easily destroy the scanty protective cover and increase the vulnerability of the environment.

Industrial activity Handling hazardous substances is a potential focus of pollution, both directly and indirectly, including air pollution.

Agricultural and forestry activities The two agricultural activities that have the greatest impact on karst waters are the storage and movement of agricultural residues (manure, slurry and other effluents) and the application of fertilizers and pesticides, which cause problems with nitrates and microorganisms, and with toxic substances.

Deforestation has the greatest indirect impact of all human activities on karst. It has left its mark since the Mesolithic, and continued during Greek and Roman time to the present day in Chap. 1 [2]. The disappearance of trees promotes the loss of soil and the capacity for surface retention. The construction of roads, the mechanical equipment and its maintenance (lubricants, oils) and the use of pesticides are all dangers.

7.4.2.2 Control Measures

General aspects Protection of the karst environment involves taking preventive measures to guarantee the natural quality of the water. Karst water supplies many villages and a range of services, including fish farms and recreational activities. This means that some restrictions and even prohibitions must be introduced that may clash with private interests and generate conflict. The priorities need to be defined by the state through relevant laws.

The objectives of control measures are to prevent or prevent toxic substances from reaching the aquifer waters and also to ensure a good quantity. The risks associated with the production, transport, storage and application of toxic substances must be eliminated. Precise control decisions need to be discussed by hydrogeologists and other specialists, complemented by appropriate research. There is a problem specific to liquid wastewater, which spreads easily and infiltrates quickly, so surface run-off must be taken into account. The content of the discharge requires prior stabilization, even to the extent of generating an insoluble form, in the case of fertilizer.

Categories for the polluting potential of the waste in question may well have to be established. The prime category would include industrial wastewater and leachate from landfill, which requires pre-treatment prior to discharge. A second category could include run-off from roofs or roads in the event of accidents, and related measures and precautions.

Exploitation control The intensive exploitation of aquifers can have strongly negative consequences not only on the quantity but also on the quality of the water, hence the need to plan adequately the maximum permitted exploitation that can guarantee the balance of the system. However, one of the great difficulties of such a decision is to ascertain with precision the amount of the recharge, subject to many uncertainties. The quantity pumped and, in addition, the exact distribution of the descents must be studied to prevent the springs from drying out. Hence, there is a need to establish an adequate piezometric observation network, not only over the control and monitoring of wells but the dangers of karst collapse induced by overpumping. The effects must be anticipated, given the immense difficulty of recovering a polluted karst aquifer (Olkuscz in Chap. 1 [11, 22]).

Coastal karst aquifers are a special case, in view of the risk of marine intrusion. Mathematical simulation models, with the appropriate data, allow for a sufficient range of responses by aquifers to certain exploitation policies to foresee undesirable outcomes, even including the possibility of planning artificial recharge.

Control of infrastructure development The great vulnerability of the karst environment means that the areas covering aquifers are spatially sensitive to activities and land use. Some of the possible dangers and risks are:

- Urban settlements, with the associated waste products, even for low population density (liquids, solids, slurry, etc.).
- Tourism development: The problem is similar to that of urbanization, although with a strong tendency to be sited in more singular and sensitive areas, with a possible increase in risk (wastewater, increase in traffic, etc.).
- Industry: There is greater risk in the case of industries involving hazardous substances.
- Agriculture: waste storage, treatment, etc.
- Traffic systems: Pollutant emissions into the environment, such as lead, organic material, salts in freezing weather, plus accidents involving hydrocarbons and other dangerous substances.
- Solid urban waste landfills and others: Leachate is the most dangerous.
- The demand for space for such dangerous land uses means that measures have to be taken to conserve sensitive areas. The recommended protective measures are shown in Table 7.15.

Industrial pollution control Potential conflicts of interest between industry and karst are especially difficult, for the following reasons:

- the industrial activities involved are, in most cases, of fundamental importance to the region's economy
- most karst areas are mountainous, with limited economic resources; restricting industrial activities can be perceived as a frontal attack on development
- the reduction of pollutant emissions has a cost and involves considerable investment
- in the case of mining, the extraction of a non-renewable material is often in competition with water sustainability
- industrial leaders who are not very sensitive to the environment often have a strong influence on decision-makers.

Table 7.16 summarizes the main control measures specifically for karst regions.

Control of agricultural and forestry practices Some of the factors that make agricultural activities difficult to control are:

- their locality: agricultural activities are the most affected, in many regions
- farmers are traditionally very independent

Table 7.15 Control measures for different activities

Activity	Control measure
Urbanization and Industries	Land use planning standards and guidelines for preserving the environment – detailed construction rules to ensure safe buildings, waterproof tanks and storage, rules for the installation of drainage systems for housing and roads – prohibition of disposal of contaminated untreated groundwater guidelines for safe sewerage – treatment of polluted water – appropriate quality requirements for slurry from wastewater treatment plants in order to make it usable as fertilizer in agriculture – restriction and prohibition of particularly hazardous landfills
Settlements and housing	Adequate sewage system, safe ponds for diesel and other dangerous liquids – prohibited in the vicinity of those morphological features indicating rapid infiltration or particularly vulnerable zones – prohibited near springs and wells
Roads and railways	Adequate planning and construction – prohibited near springs and wells – use non-hazardous and non-toxic construction materials – limit the use of salt (roads) and herbicides (railways) – restrictions on the destruction of the protective cover – restrictions on artificial drainage of karst aquifers (e.g. tunnels)
Airports/aerodromes	adequate location, planning and construction – safe construction of fuel tanks – construction material that does not endanger the quality of the water – restricted use of salts
Campsites	Adequate location – prohibited near springs and wells – prohibited near certain morphological features indicative of rapid infiltration
Golf courses	Adequate location – prohibited near springs and wells – prohibited near certain morphological features indicative of rapid infiltration
Cemeteries	Adequate location with deep water level and karst aquifer – prohibited near springs and wells – prohibited near certain morphological features indicative of rapid infiltration

- farmers, as a collective, have considerable political influence in many countries
- agriculture is heavily dependent on chemicals for crop improvement.

The new EU reforms included in the Community Agricultural Policy (CAP), which seeks to achieve the best possible agricultural practices in a progressive manner, must be incorporated into all of this. An example is the Nitrates Directive. Table 7.17 lists possible controls for these activities.

Table 7.16 Main industrial activities and some possible control measures

Activity	Control measure
Industrial plant	
Site selection	Industrial plants are prohibited next to wells and springs or in vulnerable areas
Program of operations	Possible measures to prevent pollution – suitable locations for the handling of liquids or solids likely to contaminate water – periodic checks to identify possible leaks – action programmes for accidents involving hydrocarbons or other pollutants – adequate containment system in case of leakage or spillage – water quality control in the emission area possible
Emission of airborne pollutants	Possible preventive measures – control of smoke emission by means of suitable filters – groundwater protection standards compatible with air protection standards
Mining, quarries and gravel pits	
Conditions	Possible measures – appropriate methods of operation – maintenance of a sufficient depth of cover above the water table – adequate drainage of the site, cleaning and drainage of rainwater and wastewater – suitable places for decanting and storing hydrocarbons and other polluting liquids – suitable places for parking and maintenance of vehicles, as well as for the storage of building materials and explosives – action programmes in the event of accidents involving oil and other pollutants – filling of abandoned excavations with suitable materials with special care – monitoring of water quality in the area of influence of operations Prohibited activities in the following situations: – in the vicinity of wells and springs, vulnerable areas and morphological environments indicative of rapid infiltration – in areas of shallow water table in exploitable aquifers
Disposal of liquid and solid waste	
Collection, conditioning and storage	Possible measures: – adequate treatment of waste in a stable form (e.g. inert, insoluble form, etc.) – adequate collection system and adequate storage place
Disposal	Possible measures: – it is generally prohibited to pour liquid waste into surface water or to inject it into geological formations – suitable site; appropriate conditions will have to be demonstrated by a study to prevent infiltration of contaminants; the environment has to meet specific standards with respect to permeability, homogeneity and topography – adequate site drainage, collection and treatment of contaminated leachate – monitoring of water quality in the area of influence of the site

(continued)

Table 7.16 (continued)

Activity	Control measure
Disposal of wastewater	
	Prohibited activity: – proximity of wells and springs – in vulnerable areas and morphological environments indicative of rapid infiltration
Collection, treatment and storage	Possible measures: – correct collection system, adequate storage and containment – periodic monitoring of sewer system leaks – wastewater treatment; must be treated in a treatment plant before disposal – special treatment of liquids such as mineral oils, solvents and pesticides
Disposal	The disposal of treated or unpolluted wastewater is possible if it does not endanger the quality of groundwater; possible preventive measures are: For treated or uncontaminated wastewater: – discharge into surface water – spreading or throwing on biologically active soil – spreading or throwing on gravel or sand with suitable filtering properties For uncontaminated wastewater: – infiltration The following measures are generally prohibited: – disposal of untreated industrial wastewater or other polluting liquids such as mineral oils, solvents and pesticides on surface or groundwater (e.g. injection, infiltration or spraying) – removal of wastewater in the vicinity of wells and springs and vulnerable areas and morphological environments indicative of rapid infiltration – discharge of liquids such as mineral oils, solvents or pesticides

7.4.3 Developing Protection Procedures

7.4.3.1 General Information

Karst water in its natural state is an excellent quality resource in places far from spring water or where it is very scarce; hence, it is correctly regarded as a drinking water supply. On the other hand, the karst environment is usually highly vulnerable, hence the need for protection and control. As prevention is always better than cure, it is necessary to establish an applicable and effective protection system based on the establishment of protection zones, on the development of appropriate land use and on the correct handling of toxic substances—and the whole dependent on an functioning monitoring system.

In the natural environment, the common good must prevail over vested interests, especially when it comes to water. Hence, in a conflict that may arise between the legitimate interests of the owners and the protection of the water, it should be the

Table 7.17 Possible control measures in agricultural and forestry activities

Activity	Control measure
Manure and animal slurry	
Collection and storage	Adequate collection system, correct storage and containment: – generally prohibited on outcrops of carbonated rocks – prohibited near morphological traits indicating rapid infiltration or in sinks or losing streams – prohibited in the vicinity of wells or springs – where the soil cover is thin, construction methods should minimize its elimination.
Application	Appropriate application policy: – no spreading on bare limestone, where the soil cover is thin, near morphological features that indicate rapid infiltration or near losing sinks or streams or other surface waters. – the quantity applied should cover the needs of the crop, but no more – no spreading after a rainy period – no spreading on snow or frozen ground – preferably, spread early in the growing season – reduce the quantity applied – reduce the amount applied, harvest in the autumn and convert arable areas to pasture, in areas where high nitrate content poses problems
Cloudy water and fodder effluent	
Collection and storage	Minimize production, proper farm design, correct collection system: – minimize the use of water for cleaning pens and dairies – minimize the areas of uncovered pens for animals – construct gutters to prevent clean water from entering stabling areas – use high-quality construction materials
Disposal	Widespread, to facilitate soil attenuation
Inorganic fertilizers	
Storage	Correct location, construction and operation of storage facilities: – prohibited in the vicinity of wells and springs – prevent spills in vulnerable areas, install containment systems
Application	Appropriate application policy: – spread quantities that meet the needs of the crop, no more – no application before or during rainy periods – no application on snow or frozen ground – preferably apply nitrogen at the beginning of the growing season – reduce the quantities applied in areas where high nitrate content poses problems – harvest in the autumn – convert arable land into pasture – practise environment-friendly agriculture
Pesticides	
Storage	Correct location, construction and operation of storage facilities: – prohibited in the vicinity of wells and springs – use adequate containment systems in vulnerable areas
Application	Appropriate application policy: – use recommended dose and method of application – use biodegradable, rapidly degradable and/or non-mobilizable pesticides

(continued)

Table 7.17 (continued)

Activity	Control measure
	– avoid during unfavourable weather conditions – prohibited near wells and springs – prohibited near morphological features indicative of rapid infiltration – in forested areas, take extreme precautions
Intensive use	
Garden centres	Adequate storage facilities for fertilizers and pesticides, and correct dosages Prohibited in the vicinity of wells and springs or near morphological features indicative of rapid infiltration or in vulnerable areas
Orchards	Apply correct quantities of fertilizers and pesticides
Deforestation	Reforestation Safety margin in the vicinity of geomorphological features indicative of rapid infiltration or close to sinkholes or losing streams
Intensive forestry	Restrictions and correct applications of herbicides and insecticides
Overgrazing	Keep the number of animals to a sustainable level or compatible with sustainable use Use access controls in areas subject to erosion

latter that should prevail. The interaction between planners and hydrogeologists is vital to the optimization of karst water. Based on knowledge of the time that it takes the water to pass through and the ability of the environment to cleanse itself, more restrictive protection zones may need to be established in the areas of greatest vulnerability. The usual criterion applied to land with an intergranular porosity medium, based on distance to the polluting focus, is not applicable in karst, in principle, due to its great heterogeneity. The ducts have very short transit times and a low self-purification power. Instead, the intensity of the karstification should be a preferred criterion when establishing protection zoning. This zoning should frequently take into account that there may be intergranular porous aquifers or fissure media in the area and make policies that are consistent with all this information.

The general outline of protection will take into account the schemes devised by various countries, adapted to the needs of other countries and based on the concepts of pollution risk, risk assessment and risk management as a whole.

7.4.3.2 Risk and Risk Management: Programme for Protection

The risk of groundwater contamination depends on three factors: the hazard from a potentially polluting activity; the consequences of a polluting event; and the likelihood of contamination.

Risk management is based on the analysis of these elements, followed by a response to the risk through the implementation of preventive measures that minimize the consequences and probability of the contaminating event occurring.

The *hazard* depends on the potential pollutant load. The economic value of groundwater normally associated with the category of the aquifer—large, small or poor—or its proximity to a source makes it possible to assess the consequences if the polluting event does arise. The probability of contamination depends on the natural vulnerability of the water. Preventive measures may include the control of land use practices, which should be in line with the development of low-risk areas; codes that take into account the vulnerability and value of groundwater; lining of waste dumps; implementation of control and monitoring networks; and a series of specific operating practices. Consistent with the foregoing, we can conclude that the assessment of the risk of groundwater contamination is somewhat complex and must consider geological and hydrogeological factors, together with factors relating to the potentially polluting activity. The former are: (a) the relative importance or value of groundwater; and (b) vulnerability to pollution. The factors related to the polluting potential of the activity are: (a) the polluting load; and (b) preventive measures.

The groundwater protection plan should both integrate these factors to focus attention on high-risk areas and activities and provide a logical structure in which pollution control measures should be imposed.

The two main components of the protection plan are: (a) a zoning map, or maps, of the land surface, often called a groundwater protection map, which integrates the hydrogeological elements of the risk; and (b) a code of practice for potentially polluting activities, existing or new which integrates the polluting load and the elements of pollution risk control.

7.4.3.3 Zoning of the Medium

The protection plan is drawn on the zoning map to create differentiated protection areas and thus impose the required regulation. Some parameters for zones require little hydrogeological data; some even establish radii arbitrarily around water sources. Others, on the other hand, are highly sophisticated, based on involved data manipulations. Although the simplest methods may be a first step, ideally the zoning should always be based on hydrogeological data of as diverse a range as possible.

The three fundamental elements of surface zoning are:

- Protecting the areas around groundwater sources; this is *source protection.*
- Subdividing the area on the basis of the value of the resource or the aquifer's category; this is *resource protection.*
- Categorizing the entire area according to its groundwater's vulnerability to pollution.

Source protection is of particular importance. The simplest procedure is to establish a circular zone around the source, with decreasing protection with increasing distance. The disadvantage is that this ignores features such as highly vulnerable chasms and sinks, while overprotecting places situated downstream and leaving unprotected the upstream sector, in the direction of the groundwater flow.

A suggested approach for karst sources is to have two main protection zones: the location of the spring or source; and the catchment, divided in categories according to its water's vulnerability to pollution. The number of categories may vary, but there are usually at least three. In addition, the information on which that categorization is based may vary. It should, at least, include sumps, rapid recharge chasms, the geological nature of the materials above the water table and all known karst ducts. Vulnerability maps are a basic risk assessment tool for karst springs.

One of the greatest difficulties in the karst environment is getting to know exactly the real limits of the catchment area of a spring. Experiments with tracers, precise balances and isopiece maps would be possible procedures, although none is conclusive, for the simple reason that such parameters vary according to the hydrodynamic conditions of the moment. Mathematical simulation models, with their apparent high precision, can be very misleading, although they contribute a great deal to characterizing flow systems. For this reason, it is always recommended to err on the side of safety, covering a larger area than a smaller area.

Resource protection is usually concerned with the value of the resource. Note that this is a very commercial view of water, and this in itself is an advantage as it covers a wide range of environmental services, whereas the economic value is not always in the common good. Three categories are usually differentiated:

(a) **larger aquifers**, responsible for large supplies of drinking water
(b) **minor aquifers**, without relevance to the water supply owing to their characteristics; and
(c) **poor or non-aquiferous aquifers**.

Karst aquifers are often in the first category.

7.4.3.4 Code of Good Practice

This code shows the extent of acceptability of potentially polluting activities for each zone and sub-area, and describes the recommended controls for activities, both existing and planned. The level of response depends on the various risk elements already indicated. At least four levels of response, R, can be considered regarding the risk of a potentially polluting activity:

- RI: acceptable
- R2a, b, c: acceptable in principle, subject to conditions a, b, c, etc. The number and content of such notes may vary for each zone and subzone and for each activity.
- R3m, n, o: unacceptable, in principle; some exceptions may be made subject to the conditions indicated in footnotes m, n, o, etc.
- R4: unacceptable.

The final stage in a groundwater protection plan is the integration of protection zones and the code of practice. Table 7.18 shows the foregoing diagrammatically, in the form of a matrix.

Table 7.18 Assignment of response R levels to the protection zones

Vulnerability	Source protection			Protection of the resource				
	Source	Interior	Exterior	Main aquifer		Intermediate aquifer	Poor aquifer	
Extreme	R4	R4	R4	R4	R4	R3m	R2m	↓
High	R4	R4	R4	R4	R3m	R3n	R2b	↓
Moderate	R4	R4	R3m	R3m	R3n	R2b	R2m	↓
Low	R4	R3m	R3m	R3mn	R2m	R2a	R1	↓
	→	→	→	→	→	→	→	

The matrix integrates the geological/hydrogeological aspects and those of the contaminant load into the risk assessment. In general, the arrows →↓ indicate a decreasing risk, where ↓ shows the decreased probability of contamination and → a decreased consequence. The risk load is indicated by the activity in the column heading. The response to the risk of groundwater contamination is given by the category assigned to the response for each sub-area and by the site investigations and/or controls and/or protective measures described in notes *a*, *b*, *c*, *m* and *n*.

EU directives have gradually introduced new ideas that have been responded to by the scientific field. The doctoral thesis by Jiménez Madrid in [31], 'Estudio metodológico para el establecimiento de zonas de salvaguarda de masas de agua subterránea en aquíferos carbonatados utilizados para consumo humana', is notable. *Implementation of the Water Framework Directive* and several of its studies [32–36] are recommended for consultation.

7.5 Further Reading

Al-Bassam, A.M., Khalil, A.R. 2012. DurovPwin: A new version to plot the expanded Durov diagram for hydrochemical data analysis. *Comput. Geosci.* 42: 1–6.

Khadra, W.M., Stuyfzand, P.J., van Breukelen, B.M. 2017. Hydrochemical effects of saltwater intrusion in a limestone and dolomitic limestone aquifer in Lebanon. *Applied Geochemistry,* 79: 36–51.

Liu, Y., Jiao, J.J., Liang, W., Kuang, X. 2017. Hydrogeochemical characteristics in coastal groundwater mixing zone. *Applied Geochemistry*, 85: 49–60.

Moore, P.J., Martin, J.B., Screaton, E.J. 2009. Geochemical and statistical evidence of recharge, mixing, and controls on spring discharge in an eogenetic karst aquifer. *Journal of Hydrology,* 376: 443–455.

Parkhurst, D.L., Appelo, C.A.J. 2013. Description of input and examples for PHREEQC version 3. A computer program for speciation, batch-reaction, one-dimensional transport, and inverse geochemical calculations: U.S. Geological Survey Techniques and Methods, Book 6, Chap. A43, 497 p., available at http://pubs.usgs.gov/tm/06/a43/

Pulido-Leboeuf, P. 2004. Seawater intrusion and associated processes in a small coastal complex aquifer (Castell de Ferro, Spain). *Applied Geochemistry*, 19: 1517–1527.

Yechieli, Y., Yokochi, R., Zilberbrand, M., (…), Livshitz, Y., Burg, A. 2019. Recent seawater intrusion into deep aquifer determined by the radioactive noble-gas isotopes 81Kr and 39Ar. *Earth and Planetary Science Letters*, 507: 21–29.

Zwahlen, F. 2003. *COST Action 620. Vulnerability and risk mapping for the protection of carbonate (karst) aquifers—final report*. European Commission.

7.6 Short Questions

1. Under normal conditions, what are the major anionic components in the waters of a karst aquifer?
2. Majority cationic components in karst waters.
3. Can correlation and spectral analysis be applied to hydrogeochemical data?
4. How would large fluctuations in electrical conductivity in the waters of the main source of a karst aquifer be interpreted?
5. Why do frequency distribution analyses give the degree of karstification? Give reasons for your answer.
6. To what extent are lines of equal concentration useful in the study of karst aquifers?
7. Is there a vertical gradation in salt content in complex karst aquifers? Give reasons for your answer.
8. Why do hydrogeochemical parameters vary greatly in complex karst aquifers?
9. Why is the strip where fresh and saltwater mix in coastal aquifers usually in a stretch of great karstification?
10. List the main processes that stable isotopic data can identify in large and complex karst aquifers.
11. List the main processes that radioisotope data can identify in large and complex karst aquifers.
12. What are the usual chemical characteristics of water in evaporite aquifers?
13. What indicates the presence of sodium chloride facies water in aquifers apparently formed by gypsum?
14. Why are karst aquifers often considered highly vulnerable to contamination?
15. Should all anthropogenic activities be prohibited on karst land? Give reasons for your answer.

7.7 Personal Work

1. Problems of coastal karst aquifers.
2. Agricultural pollution in karst aquifers.
3. Dangers and risks associated with karst waters.
4. Methods of mapping karst aquifers' vulnerability to contamination.

References

1. Hem, J. D. (1985). *Study and interpretation of the chemical characteristics of natural water* (4th ed., p. 263). U.S. Geological Survey Water-Supply Paper 2254: Washington.
2. Catalán, J. (1969). *Química del agua* (p. 423). Madrid: Ed. Blume.
3. Custodio, E. (1986). Hidrogeoquímica del karst. *Jorn. Karst Euskadi* (Vol. 2, pp. 131–179). San Sebastián.
4. Custodio, E. (2001). Hidrogeoquímica. In E. Custodio, & M. R. Llamas (Eds.), *Hidrología Subterránea* (Sec. 10, pp. 1005–1091). Barcelona: Ed. Omega.
5. Custodio, E. (2001). Principios básicos de Química y radioquímica de aguas subterráneas. In E. Custodio, & M. R. Llamas (Eds.), *Hidrología Subterránea* (Sec. 4, pp. 177–245). Barcelona: Ed. Omega.
6. Benavente, J., Moral, F., Vallejos, A., & Pulido-Bosch, A. (2004). Hidroquímica de acuíferos kársticos. In *Investigaciones en sistemas kársticos españoles* (Vol. 12, pp. 139–159). IGME, Serie Hidrogeología y Aguas Subterráneas.
7. Parkhurst, D. L., & Appelo, C. A. J. (1999). *User's guide to PHREEQC (Version 2): A computer program for speciation, batch-reaction, one-dimensional transport, and inverse geochemical calculations.* U.S. Geological Survey, Water-Resources Investigations Report 99–4259, p. 312.
8. Pulido-Bosch, A. (1975). Los manantiales salinos de la Sierra de Mustalla. *Jorn. Min. Met., III*, 117–128. (Bilbao).
9. Ballesteros Navarro, B. J., Domínguez Sánchez, J. A., Díaz-Losada E., & García Menéndez O. (2009). Zonas húmedas mediterráneas y acuíferos asociados. Condicionantes hidrogeológicos del Marjal de Pego-Oliva (Alicante-Valencia) *Boletín Geológico y Minero, 120*(3), 459–478.
10. Moral, S., Pulido-Bosch, A., & Valenzuela, P. (1984). Aplicación de los análisis "cluster" al estudio de características físico–químicas de aguas subterráneas. *Estudios Geológicos, 40*, 193–200.
11. López Chicano, M., & Pulido-Bosch, A. (1995). Los manantiales termominerales de Salar (Granada). Un sistema de flujo profundo ligado esencialmente a la descarga de sierra Gorda. *Geogaceta, 18*, 138–141.
12. Vallejos, A. (2001). *Caracterización hidrogeoquímica de la recarga de los acuíferos del Campo de Dalías a partir de la Sierra de Gádor (Almería)* (p. 242). University of Almeria–IEA.
13. Mazor, E., Jaffé, F.C., Fluck, J. & Dubois, J.D. (1986). Tritium corrected ^{14}C and atmospheric noble gas corrected ^{4}He applied to deduce ages of mixed groundwaters: Examples from the Baden region, Switzerland. *Geochimica et Cosmochimica Acta, 50*(8), 1611–1618.
14. Zuber, A., Weise, S. M., Osenbrück, K., Grabczak, J., & Ciezkowski, W. (1995). Age and recharge area of thermal waters in Ladek Spa (Sudeten, Poland) deduced from environmental isotope and noble gas data. *Journal of Hydrology, 167*, 327–349.
15. Wood, W. W., & Sanford, W. E. (1995). Chemical and isotopic methods for quantifying groundwater recharge in a regional, semiarid environment. *Ground Water, 33*(3), 458–468.
16. Araguás, L. (1991). *Adquisición de los contenidos isotópicos (18O y D) de las aguas subterráneas: variaciones en la atmósfera y en la zona no saturada del suelo* (p. 286). Thesis, University of Madrid.
17. Vallejos, A., Pulido-Bosch, A., Martin-Rosales, W., & Calvache, M. L. (1997). Contribution of Environmental Isotopes to the knowledge of complex Hydrologic Systems. A case study: Sierra de Gador (SE Spain). *Earth Surface Processes and landforms, 22*, 1157–1168.
18. Fritz, P., Drimmie, R. J., Frape, S. K. & O'Shea, K. (1987). The isotopic composition of precipitation and groundwater in Canada. In *Proceedings of a Symposium on Isotope Techniques in Water Resources Development* (pp. 539–550), IAEA.
19. Rindsberger, M., Jaffe, Sh., Rahamin, Sh., & Gat, J. R. (1990). Patterns of isotopic composition of precipitation in time and space: data from the Israeli storm water collection program. *Tellus, 42B*, 263–271.

20. Pulido–Bosch, A. (1985). Groundwater mining In the aquifer Sierra de Crevillente and its surroundings (Alicante, Spain). In *Hydrogeology in the service of man* (pp. 142–149). Memories 18th Congr. Cambridge: IAH.

21. Pulido-Bosch, A. (1988). The overploitation of certain karstic aquifers of Alicante (eastern Spain). *Bulletin du Centre d'Hydrogéologie Univ. Neuchâtel, 8*, 49–60.

22. García Hemández, M., Molina, J.M., Ruiz–Ortiz, P.A., & Vera, J.A. (1989). Wedging and sigmoidal geometry in red pelagic Jurassic limestone of the Sierra de Redot (Alicante province). *Congreso Geología. España, 1*, 83–86.

23. Azema, J. (1977). *Geological study of the external zones of tbe Betic Cordilleras in the provinces of Alicante and Murcia (Spain), VI* (p. 395). Doctoral thesis. Paris.

24. Arikan, A. (1988). Basin revision of WATEQF for IBM personal computer. *Ground Water, 26*(2), 222–227.

25. Tulipano, L., & Fidelibus, M. D. (1991). Modern orientation on the karstic hydrology: Impact on problems of groundwater protection into carbonate aquifers. Experience from the Apulia Region. *Quaderni del Dipartimento di Geografia, 13*, 383–398.

26. Plummer, L.N., & Busenberg, E. (1982). The solubilities of calcite, aragonite, and raterite in Co –HO solutions-between O and 90 °C *Cosmochem. Acta, 46*, 1011–1040.

27. Langmuir, D. (1984). Physical and chemical characteristics of carbonate water. *Guide to the Hydrology of Carbonate Rocks. Studies and Reports in Hydrology, 41*, 264–265.

28. Calaforra, J. M., & Pulido-Bosch, A. (1993). The hydrogeochemistry and morphology of the Triassic gypsum in the Salinas-Fuente Camacho area (Granada). *Some Spanish karstic aquifers* (pp. 67–83). Granada, Spain: University of Granada.

29. Fagundo, J. R. (1996). *Hidrogeoquímica del karst en climas extremos* (p. 212). Granada: University of Granada.

30. Parkhurst, D. L., Plummer, L. N. Thorstendson, D.C. (1982). *BALANCE–A computer program for calculating mass transfer for geochemical reactions in groundwater* (p. 19). U.S.G.S: Water Resources Investigations.

31. Jiménez-Madrid, A. (2011). *Estudio metodológico para el establecimiento de zonas de salvaguarda de masas de agua subterránea en acuíferos carbonatados utilizados para consumo humano. Aplicación de la Directiva Marco del Agua* (p. 436). Thesis, Doct., University of Malaga.

32. Jiménez–Madrid, A. Martínez–Navarrete, C., & Carrasco, F. (2010). Groundwater risk intensity assessment. Application to carbonate aquifers of the western mediterranean (Southern Spain). *Geodinámica Acta, 23*(1–3), 101–111.

33. Jiménez–Madrid, A., Carrasco, F., Martínez, C., & Vernoux, J. F. (2011). Comparative analysis of intrinsic groundwater vulnerability assessment methods for carbonate aquifers. *Quarterly Journal of Engineering Geology and Hydrogeology, 44*, 361–371.

34. Martínez–Navarrete, C., Jiménez–Madrid, A., Sánchez–Navarro, I., Carrasco F, & Moreno–Merino, L, (2011). Conceptual framework for protecting groundwater quality. *Water Resurses Dev, 27*(1), 219–235.

35. Jiménez–Madrid, A., Carrasco, F., & Martínez–Navarrete, C. (2012). Protection of groundwater intended for human consumption: a proposed methodology for defining safeguard zones. *Environmental Earth Sciences, 65*, 2391–2406.

36. Jiménez–Madrid, A., Carrasco, F., Martínez, C., & Gogu, R. C. (2013). DRISTPI, a new groundwater vulnerability mapping method for use in karstic and non–karstic aquifers. *Quaterly Journal of Engineering Geology and Hydrogeology, 46*, 245–255.

Exploration and Exploitation

8

8.1 Glossary

Hydrogeological cartography is a basic exploration tool. International works have been published on the subject. *Fracking analysis*, *aquifer point inventory*, *hydrochemical data*, *geophysical methods*, and *tracers* are useful research aids for the karst environment.

Borehole *logs*, *isothermal maps and conductivity profiles* can be very useful when there are several mechanical boreholes.

Infiltration into karst aquifers is usually higher than into intergranular porosity media, and *transit speeds* are also higher, also the *average storage* is usually much *lower*. Consequently, the regulation of a karst system is more complex than that of a detrital system.

The estimation of potential recharge using an ecohydrological model is based on the hydrological equilibrium hypothesis, which suggests that vegetation evapotranspiration balances water use and water stress and that vegetation in a region unaffected by human influence is in equilibrium with the available water. The Normalized Density Vegetation Index (NDVI) is a very useful parameter.

Chloride balance methods for estimating potential recharge in transit, and certain *numerical models* are also very useful.

Artificial tracers are used in karst exploration, although the essential objective is usually to determine the connection between two points and the *transit times*. They are an essential tool in studies related to the *propagation of pollutants*.

Although it cannot be considered an exploration tool, the use of *tracers* is essential in drawing up an adequate policy for the rational exploitation of a karst aquifer. It enables the possible responses, effects, total pumpable volume and consequences for the spring's discharge to be determined.

© Springer Nature Switzerland AG 2021
A. Pulido-Bosch, *Principles of Karst Hydrogeology*, Springer Textbooks in Earth Sciences, Geography and Environment,
https://doi.org/10.1007/978-3-030-55370-8_8

The main methods of water extraction are: the diversion of a spring; excavating a gallery into a spring; pumping from places of direct access (*wells and similar*); making a reservoir in a spring; a reservoir complemented by horizontal drilling and a capping; vertically drilling downstream of the spring; doing this horizontally and capping it; excavating a gallery downstream of the spring; excavating a gallery with boreholes within it; excavating a gallery and drilling both vertically and horizontally; wells and galleries with boreholes under a spring; well and galleries under a spring, with sideways extensions and a capping; m: underground dams and upstream boreholes; waterproofing submarine karst conduits; vertical wells and horizontal galleries at sea level (coastal aquifers); drilling boreholes upstream of the spring.

The most usual way of *regulating a karst aquifer* is to drill a *borehole*, with characteristics dependent on the intended objective. The recommended procedure is *rotary percussion drilling*.

Acidification is the best way to achieve the *stimulation and development of boreholes* that exploit carbonate aquifers (limestone and dolomite and sandstone and conglomerate of carbonate cement). *HCl* is the most commonly used acid.

The main *direct impacts* of intensive exploitation of aquifers are: a *decrease in piezometric level*; an *increase in the cost* of exploitation; a *deterioration in water quality*; the need to *abandon boreholes*; alterations to the *fluvial regime*; changes in any *wetlands* associated with the aquifer; and *legal complications* with third parties.

The *indirect impacts* are: the *salinization* of the soil; progressive *desertification*; initiation of *collapse*; changes in the *soil's physical properties*; and *induced contamination* at distance due to intensive pumping.

8.2 Background

The basis of hydrogeological research and prospecting in karst environments has many points in common with that in porous media. However, it must be borne in mind that karst is a generally heterogeneous and anisotropic medium, so the more restricted are the observations, the more marked are the results.

Hydrogeological mapping is the basic tool. It should be as near as possible to geological mapping but using hydrogeological parameters. It involves marking out those lithological sets with similar hydrogeological behaviour. This is why, in many cases, it is of no use to separate out the limestone from the dolomite. Small sections of a dissimilar hydrogeological nature, unless they have a guiding value, are likewise of insufficient importance—20 m of marl in 500 m or 300 m of limestone has no influence at the scale of an entire system that has been folded and fractured. It is a function of scale: the smaller the scale of the work, the more detail is required.

An international publication by UNESCO (1970) has graphic representations of the hydrogeological aspects of geology, extended by later works. In general, geological data of many kinds are usable and have practical application if we bear in mind that most karst aquifers that are studied are in regions that have been folded, so their structure is crucial. Karst features of hydrogeological interest must be marked on these maps, especially those forms that promote immediate absorption, and these will influence the infiltration of water.

Fracturing analysis is an auxiliary study technique, as examined in Chap. 1. The other key element is an *inventory of aquifer points.* Used in conjunction with mapping, this gives a precise and qualitative idea of an aquifer's potential. A series of data is associated with technique. Statistical analysis of the values obtained, performance data for pumping tests and analysis of graphs provide the basic information to define a system's characteristics and potential. The flow of a spring and/or well and an analysis of its altitude illustrate their relationship to the whole system, and will reveal if it is isolated and/or perched. Ultimately, we can trace *pseudo water table contours* and estimate the gradients that show lower values, and thus identify the more permeable zones.

The hydrogeological base map, of critical importance in planning water resource exploitation, can be accompanied by diagrams showing *hydrochemical data,* and full details of the fluvial network, geographical data and topography must always be provided. Small maximum, minimum and average rainfall planes are also available in a diagrammatic form. Hydrogeological cross-sections are crucial (on large-scale geological maps showing details of the piezometric level, springs, wells and boreholes). The analysis of hydrographs is also a most important research technique.

The most widely used geophysical methods [1] in hydrogeological research are electrical, seismic, gravimetric and micro-gravimetric. Within the electric group are electric soundings, tomography, electric trails, multidirectional soundings, *mise à la masse,* electromagnetic and radar, besides using borehole logs. In no case does any geophysical method determine a system's productivity. Rather, it seeks either continuities or discontinuities in the upwelling system and fractures in the substrate. Its use is of key importance in determining the system's geometry, especially beneath impervious surfaces.

Electrical trails and tomography make it possible to pinpoint the most fractured area in the carbonate mass and its lateral continuity. Other methods have been used with variable success, depending on the desired objective. Logs' application is generally to a framework different from the one being examined here, while *mise à la masse* establishes the groundwater flow direction and estimates its velocity using well-known procedures. Electromagnetic methods were successfully used in the Port-Miou study [2, 3], making it possible to detect a karst conduit at 50 m depth. However, the spontaneous polarization, *mise à la masse* and resistivity methods did not yield satisfactory results.

Seismic and micro-seismic methods have had mixed success in locating carbonate levels and caves, respectively. Both are usually expensive, although refraction and, above all, reflection permit great accuracy, due to the contrast in propagation patterns between a karstified and a non-karstified zone. They have been used mainly under impervious material to determine a system's depth and lateral continuity.

Micro-gravimetric methods are successfully used in the detection of caves of a certain size as a consequence of the anomalies that they create in the readings. Since the appearance of the micro-gravimeter in 1968, great precision can be achieved easily. The readings show anomalies that are of greater extent than the cavity in question, as there is a decrease in rock density around any cavity, as a result of stresses.

Apart from tracers, to which we will dedicate an entire section, there are techniques intended for specific cases, for instance *infrared thermography*. Changes in colouration show temperature change very sensitively in this method, and it has been used successfully to locate submarine springs related to karst massifs and, on land, to pinpoint areas with a shallow piezometric level, and so on. Satellite photos provide a range of data useful in hydrogeological investigations. Finally, high-precision thermal records can yield data on the operation of aquifers, the areas of mixing, the interconnection of systems, the presence of vertical flows, and so on.

Tracers [4–6] are vital tools for researching karst aquifers, and were first used by speleologists to determine the outflows of the *underground rivers* that they discovered, giving an estimate of the fastest circulation between the injection and exit points. The most traditional tracers are dyes. They are of great interest, because they make it possible to ascertain the circulation's operating mechanisms and, above all, to foresee the effects of specific pollutants. To obtain an idea of a karst system, the injection and the outlet points should be within the same aquifer, hence the paramount importance of establishing the system's geometry, especially the portions that, together, supply a given spring. The tracer is collected at the exit point, allowing a graph to be drawn (a restitution curve, showing the concentration as a function of time), analysis of which allows a series of conclusions to be formed [7].

The apparent velocity of circulation varies remarkably between a location that is positioned right on a drain (0.1–1 m/s), close to it (1×10^{-2} to 1×10^{-1}) or distant from it (1×10^{-3} m/s). This speed decreases in times of low water, and it increases with proximity to the exit point. For this reason, a single introduction of a tracer is not very informative, and it is necessary to make several. Preferential drainage zones can be determined by introducing the tracer at several points.

8.3   Exploration Methods

8.3.1   General Considerations

The system's operation and degree of karstification—provided its recharge is only from infiltration of the rainfall at the surface—can be deduced by using an overlay on the hydrograph, onto which precipitation is superimposed. In this way, when the response of a spring to rainfall is immediate, abruptly decreasing in flow when the rain stops, a highly karstified system of low regulatory power, and possibly scarce reserves, can be deduced. In this sense, a simple parameter to enable the comparison of one system to another is the Qmax/Qmin ratio: the higher it is, the more kars-tified is the system. Since the dimensions of the carbonate massif play a role in a spring's response, it is always necessary to use more advanced techniques also, such as the analysis of recession graphs or, better still, correlation analysis and simple and crossed-spectral analysis, or deconvolution, of the hydrograph over several cycles.

In Torcal-type aquifers of great inertia, considerable reserves, long memory effect, exploitation by means of boreholes can give good results by following elementary geological and hydrogeological criteria. This is the case of many of the aquifers in the Mediterranean area (Spain, Morocco, Algeria, Tunisia, Libya, and so on), which have been exploited intensely during the last forty years. On the other hand, in Aliou-type aquifers, characterized by an absence of reserves, very rapid circulation and restricted to a single a transmissive conduit, or very few conduits, exploitation is limited to searching for and abstracting water from the conduit and/or to some type of work in the spring itself.

The investigation of karst aquifers, from the point of view of quantifying the resource, determining the system's operation and/or locating a drilling area, has elements in common with the methods for the investigation of other types of aquifers. However, there are also numerous peculiarities, both in techniques and interpretation.

The Action 65 of the European COST Programme produced a document sum-marizing its work, and Chap. 3 contains aspects related to research methods. This document was based on a detailed survey by the 16 participating countries. Of the methods listed, some make it possible to characterize the karst environment and others the flow and transport phenomena, and both constitute the basis of rational management of a water system in Chap. 3 [11].

These methods listed by the borehole report can be used to obtain general knowledge about the system or, for instance, about the location of sites of exploitation. Others can be used from within the massif—soundings in caves—often as specific variants. In this case the uncertainty is probably greater, depending on the characteristics of the karst aquifer. A general conclusion is the need to complement the observations; more than one method is always essential to reduce the ambiguities that each method separately presents. Among those that provide information on the karst environment, we have morpho-structural analysis, geo-physical methods, from the surface and in boreholes, and hydrodynamic mea-surements and analysis. The last is especially used in hydrographic analysis and pumping tests, characterizing the flow and transport in the system.

The most specific aspects of the flow are: its balance, recharge and distribution of infiltration; its hydrogeochemistry and natural and artificial tracers; and its quantification by simulation. The combined application of these methods makes it possible to characterize the system in terms of its geometry and operation, available resources and the possibility of collection.

8.3.2 Morpho-Structural Analysis

Morpho-structural techniques make it possible to mark the recharge areas, system limits, divides, dissolution forms, palaeodirection of current and direction, discharge zones, favourable sites for catchment, boundary of proposed protection zones and vulnerability maps, stresses, tectonic evolution, speleogenesis, morphogenesis, and so on. The range of possibilities is extremely wide.

The tools for this type of analysis are as varied and as sophisticated as anyone could wish: topographic maps, aerial photos at various scales, infrared images—highly sensitive to temperature changes—and satellite images, among others. Classic structural analysis complements observations of outcrops, roads, caves and other points of access to the endokarst.

The data susceptible to cartographic representation and/or digitization depend on whether the scale is regional or more local. If regional, the most relevant guidelines are the main fractures, geological contacts, cappings, drainage network, rocky escarpments, 'blind' and 'dead' valleys, karstification plans, drains, chasms, caves, poljes, dolinas, alignments of exokarst forms, divides, aquifers and, thermal anomalies along the coast. At a local level, the observations are of smaller forms and those less visible at regional scale. Among others, the features that can be represented and/or measured are minor fractures, fissures, stylolites, veins, striae in the fracture planes, morphological projections, oyster-rich layers, channels and ooze.

All these data can be studied statistically, by their geostatistical spatial position or by representing graphically their spatial density to obtain information on the functional features, the tectonic history of the area [8], the relationship between these features or the past flow and current. These are often compared to the results obtained by other techniques, such as using tracers. The use of Geographic Information Systems (GIS) is a good complement [9, 10] and is an open field in karst research.

8.3.3 Geophysical Methods

8.3.3.1 General Considerations

Geophysical methods, whether airborne, surface or down boreholes, are vital tools in exploration, both in detecting discontinuities and the limits of systems, and in monitoring of tracers and, very frequently, siting boreholes or other catchment sites. Within boreholes they are tremendously useful in determining the formational parameters, physicochemical composition of the waters, the interconnection of aquifers, and the hydraulic and dimensional parameters.

The methods usually used are either electrical—using a vertical electrical survey (SEV), whether multidirectional or square, in the case of determining anisotropies—or trial pits. A pit allows us to characterize the discontinuities that will eventually be productive, although perhaps not always feasibly so. Also, there are method variants, for instance equipotential lines, such as that of a loaded body or *mise à la masse*.

SEV is a useful tool to distinguish a karstified carbonate strip from a non-karstified or a slightly karstified one. The strips show a noticeable contrast in resistivity. There is the added difficulty of fixing the electrodes when the soil has been eroded. Possibly one of the best ways to use SEVs is in the detection of carbonate material under a layer of another material of a hydrogeological nature— clays or loams. The geoelectric contrast is usually pronounced enough to be identified. Equally, this contrast can be perceived at the fresh/saltwater interface beneath a layer of substrate.

Electromagnetic methods, although in general more expensive to apply, can be much more conclusive than conventional electric methods [11], detecting the location of discontinuities. The various types of Electromagnetic Probes, VLF, Slingram and Induced Polarization, both time and frequency, that can identify lithological interfaces and large fractures well.

Seismic methods, especially reflection, are extremely conclusive, but their application is very costly. Refractive seismic tools reveal striking anomalies when there are major discontinuities in the subsoil—caverns—but give less conclusive results in most other cases, especially if the geometry of the lithological interface is highly irregular. Radar—with application principles very similar to those of seismic methods—is a technique with great potential for detecting fractures, discontinuities and even shallow caves (at less than 50 m depth). The results that can be obtained are satisfactory in both the trial pit and the speed probe mode.

Gravimetry and microgravimetry have been successfully used to locate fracture zones and shallow caves, although in real life, in addition to being costly, their application involves uncertainties that mean that they are little used. This is due to the fact that karst areas, especially mountain karst, have a very abrupt topography, so the corrections necessary to obtain the Bouguer anomaly can introduce errors that are greater than the anomaly generated by the karst feature in question.

8.3.3.2 Example of Isothermal and Isoconductivity Maps and Cross-Section Applications

Isothermal and isoconductivity soundings are used to explore karst to obtain a perspective, not only because of the information they provide on the work itself but because they make it possible to identify preferential flows, the presence of vertical feeding and discontinuities. Possibly the soundings with the highest resolution are thermometric readings. In addition to being inexpensive and simple, they have been applied with great success in Italy [12] and in the Campo de Dalias [13]. In essence, the method consists of recording water temperatures in a series of soundings with a probe capable of high precision (± 0.05 °C). Areas of preferential flow will show a negative anomaly during a feeding period, when the ambient temperature is lower than the annual average. Meanwhile, sectors with an ascending vertical feed will show a positive thermal anomaly. Although it is a very easy method, it is necessary to have enough probes of an established and identical construction and penetration to determine the three-dimensional distribution of the temperatures and, for this reason, the method's actual applicability is tightly restricted to the range of sectors where, in addition to boreholes and wells, certain places with direct access to water can serve as logging points.

8.3.4   Hydrodynamic Methods

This hydrodynamic section includes hydrogeological techniques as conventional as pumping tests—in boreholes or springs [14]—or less well-known methods such as injection, piezometric evolution and spring flows. The heterogeneity of the medium and the complexity of its operation mean that isolated data, taken without study of the correct periodicity, run the risk of providing information that can be misinterpreted [15].

Tracing isopiece graphs can be difficult when the network of blocks and ducts is taken into account. Frequently, theoretical partitions can go totally unnoticed and the resulting representation of the piezometric surface is very far from its reality. Measures of specific potential in the same vertical and temporal evolution can provide much information on the local and/or regional operation of the system [16].

Flow and/or level logs with a small time-step describe processes such as fast and/or delayed feeding, response to rainfall, contaminant propagation and blocked drain relationships. In certain cases, assuming hypotheses of doubtful validity [17], piezometric evolutions may enable estimations of the values of storage coefficients and the transmissivity of ducts, fractures and matrix.

8.3.4.1   Box 1: Isothermal Logs and Maps

Conductivity and temperature readings in boreholes—especially if these are deep—can yield invaluable information on their most relevant aspects. These include the presence of flow systems, vertically ascending or descending feeds, and fractures with preferential circulation. Remember that negative thermometric gradients characterize the recharge or feed areas, and positive gradients the discharge areas [13, 18] in Chap. 3 [5].

Figure 8.1 shows the main water points used for temperature and conductivity readings in recent years. It should be noted that the first were made with a probe 125 m long, then with one of 240 m and finally with a probe measuring 500 m, which was the one that was mostly used to establish what follows.

Figure 8.2 shows a representative record—using the Piezometer n° 1223 —to show how the temperature decreases with depth, from 20.9 °C at 74 m, 20.6 at 80 m, and a steady decrease to 19.8 °C at 200 m, then to 19.0 °C at 245 m. From a depth of 380 m, temperature remains practically constant (18.8 °C). This apparent aberrant evolution is understood by looking at the conductivity log. In fact, this last parameter shows a continuous increase in value, with an abrupt jump around 300 m. It seems, therefore, that this decrease in temperature is indicative of mixing with seawater. With the

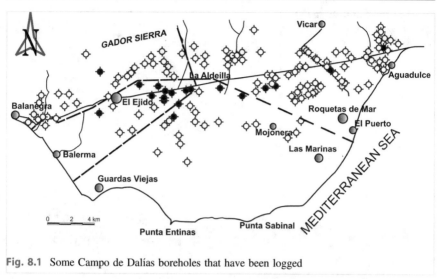

Fig. 8.1 Some Campo de Dalías boreholes that have been logged

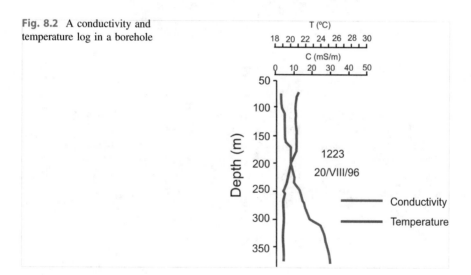

Fig. 8.2 A conductivity and temperature log in a borehole

readings plus the punctual temperature data measured in the wells in simultaneous sampling, cross-sections can be drawn up, such as those in Figs. 8.4, 8.7 and 8.8 for positions within the study area as shown in Fig. 8.3.

The $D'–D$ cross-section has a NNE–SSW orientation, ending in Guardias Viejas (Figs. 8.3 and 8.4). The water of the feeding area is cooler than 22 °C but immediately rises to 25 °C around fractures that delimit the graben immediately to the northwest of the horst of the Guardias Viejas. After this positive anomaly, the isotherms remain parallel and touching each other, with values close to 45 °C, to converge in the area of the Guardias Viejas Baths as

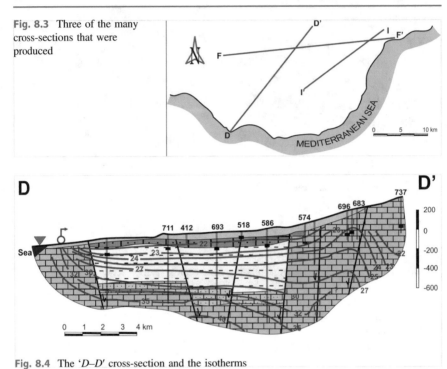

Fig. 8.3 Three of the many cross-sections that were produced

Fig. 8.4 The 'D–D' cross-section and the isotherms

the main point of discharge. The waters of the Pliocene members are at 22 °C. With the data from many cross-sections and reliable imaging, isothermal maps can be drawn for certain absolute heights, which helps us to understand the cool and warm areas and the potential to control of such flows, always dependent on the tectonics of the area (Figs. 8.5 and 8.6).

Two maps of isothermal graphs, at 100 m and 300 m below sea level, respectively, have been prepared from the aforementioned cross-sections, taking into account the fractures indicated. These horizontal cross-sections make it possible to identify, at least on an indicative basis, possible preferential flows. In the first (Fig. 8.5) the 22 °C curve delimits three 'cold water' sectors, the first to the north of El Ejido, the second in the vicinity of the Vícar wadi and the third in Aguadulce. This last reflects the marine intrusion currently affecting the main sources. The anomaly near the Vicar wadi is due to the preferential feeding water of the deep aquifer, confined under the Felix mantle. There is a certain anomaly of much lesser importance that coincides with the Balsa Nueva irrigation canal.

A positive geothermal anomaly is detected in relation to the horst of Guardias Viejas and its extension towards Aguadulce, corresponding to the ascent of the water flow towards the sea. Somewhat more curious is the positive anomaly between the Rambla del Loco and El Ejido. It is possibly an intermediate component of regional flow linked to Sierra de Gádor. In the

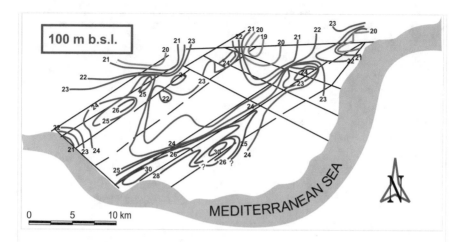

Fig. 8.5 Schematic isotherms, as deduced from readings

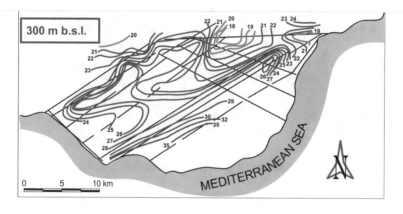

Fig. 8.6 Schematic isotherms at a depth of 300 m

second of the isothermal maps (Fig. 8.6) the cold sectors remain, but they are more localized and warm sectors have a much greater development. Given that there is only a difference of 200 m depth between the two maps, the increase in temperature is well justified.

Conductivity profiles

With the data from the thermometric profiles from the boreholes and their respective conductivity records, it is possible to draw conductivity profiles on the same basis as the thermometric cross-sections. It is possible to differentiate by colour the possible flows deduced from the isotherms and isoconductivities. The flow of cold water is shown by a blue-bordered arrow, while the arrow for the flow of hot water has a red border. The inside of the arrow can be green, if the saline content is low or black if it is high.

   The '*I*' cross-section permits the identification of a marine intrusion in
both the Gádor limestones and in the Felix and overlying materials (Fig. 8.7).
These reach 25,000 μS/cm and 10,000 μS/cm, respectively. Under the Felix
material lies Gádor (a deep aquifer), whose water has a very low saline
content as a consequence of being in the preferential flow band mentioned
above, but which already shows a warming. It cannot be ruled out that the
Miocene gypsum, crossed by deep shafts, may contribute to the salinity in
this sector (an imbalance between marine sulphates and total sulphates, in
many wells).

   The *F–F'* cross-section, running close and parallel to the edge of the Sierra
de Gádor, confirms what is exposed in the cutting intercepting it (Fig. 8.8).

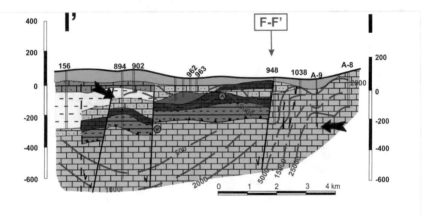

Fig. 8.7 '*I*' cross-sections and extrapolated electrical conductivity values

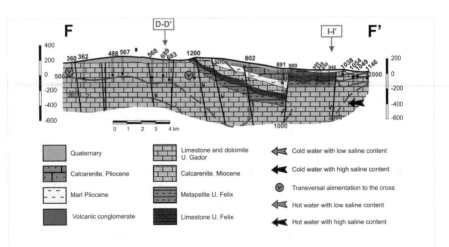

Fig. 8.8 *F–F'* cross-section showing the electrical conductivity and estimated flow types, taking
the isotherms into account

The entire western half indicates the presence of a transverse flow from the Sierra de Gádor of cool water with a low salt content, with a preferential sector under the area delimited by the Felix mantle, which is also identified by the negative geothermal anomaly. The waters increase in conductivity with proximity to the sea, where a marine intrusion has been detected, and this is very evident in the Felix materials, which have been intensively exploited for much longer.

Classic and conventional readings are also very useful, whether electronic, localized or regional, micro or macro, as well as sonic, and either naturally or artificially radioactive. Although many years have passed since its introduction in Spain, the new colour equipment, with coverage of all directions, zoom and resistance to high pressures, represents the best opportunity to recognise karst at depth, conclusively, surpassed only by direct observation in the outcrop. Correctly used, flow meters in boreholes are a tool that can provide very important information on karstification and preferential flow in mechanical boreholes and their environment [19].

8.3.5 Infiltration Recharge and Distribution

8.3.5.1 General

The fundamental objective of any hydrogeological study, especially if it is to be exploited rationally, is usually the quantification of a karst system's resources. Methods of calculating rare earth elements often encounter difficulties in their application, as there may be privileged and rapid infiltration sites linked to the cartography of the forms of absorption. In a well-defined system with a single, easily controllable spring, the calculations are simplified and the uncertainties are considerably reduced. The complications arise in areas of great geological complexity and with an abundant ground—cover vegetation—about whose limits there are always doubts. There can be variation in both space and time, and the quantification of resources is susceptible to external regulation.

Notwithstanding the foregoing, aspects of general validity for karst systems may be set out, subject to the inevitable exceptions:

- average infiltration on karst is usually higher, or much higher, than that on intergranular porosity media
- transit speeds are also higher
- medium storage is usually much lower in karst than in an intergranular porosity medium.

From all this it can be concluded that the regulation of karst systems is–normally–more complex than that of a detritic systems. The climate also influences the percentage of recharge. In semi-arid zones, the values can be very low and even nil, if rain events are of low intensity and quantity. The case of southeast Spain below is an example.

8.3.5.2 Box 2: Examples from Southeast Spain

In southeast Spain, the semi-arid climate covers an area of about 23,000 km^2 (Fig. 8.9). Average annual rainfall varies from just under 200 mm in the lower parts to 650 mm at higher altitudes. Average annual temperatures vary between 10 °C on the coast and 18.5 °C along the coastal strip. Insolation reaches 3300 h/year and the ETP 88–1200 mm/year. Estimating recharge in these areas is not easy [20, 21]. Three methods that have been applied to specific sectors of this area are shown [22].

Estimation of potential recharge using an ecohydrological model
The technique to estimate potential recharge using an ecohydrological model involves satellite data and has been applied to the Sierra de Gádor [23]. It is based on the hydrological equilibrium hypothesis, which suggests that vegetation controls its evaporation to minimize water stress and that, in a non-anthropized region, it is in equilibrium with the available water. Nemani and Running [24] developed spatially distributed algorithms to estimate annual potential recharge, I, as the difference between precipitation (P) and evapotranspiration (ET):

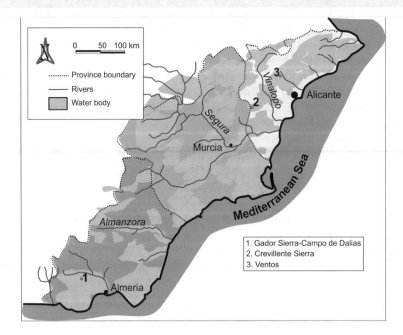

Fig. 8.9 Hydrogeological scheme of southeast Spain, showing the location of the examples

$$R = P - ET \qquad (8.1)$$

The estimate of R from Eq. 8.1 assumes that surface run-off is negligible in the long term, which can be accepted in a karst area. ET is spatially estimated using an algorithm that combines a spectral vegetation index (NDVI) and a monthly water balance model:

$$ET = (ET_{max} - ET_{min}) \frac{NDVI - NDVI_{min}}{NDVI_{max} - NDVI_{min}} + ET_{min} \qquad (8.2)$$

where NDVI is the long-term mean of the Normalized Density Vegetation Index observed at the pixel level; the pairs of values ($NDVI_{min}$, ET_{min}) and ($NDVI_{max}$, ET_{max}) are referred to those of the NDVI and ET expected for two reference conditions that are previously defined according to the mean annual precipitation (MAP) or a similar water-related index, such as the Specht Evaporative Coefficient [25]. At the pixel level, the values of $NDVI_{min}$ (bare soil condition) and $NDVI_{max}$ (vegetation cover close to its potential status, corresponding to precipitation) can be empirically adopted as the lower and maximum of the MAP-NDVI scatter plot for a sample of pixels without lateral surface and groundwater input [26]. On the other hand, the reference values of ET_{min} and ET_{max} can be estimated from the monthly water balance that integrates the seasonal average climate dynamics (P and ETP), the water retention capacity of the soil and a coefficient to represent the annual average of the evaporative conductance of vegetation [25].

With the data from the Sierra de Gádor, for the 33 basins covering a total of 552 km^2, average potential recharge values are obtained of 55, 75 and 100 hm^3/year for dry, medium and wet years, respectively. These figures are equivalent to 29, 31 and 35% of the precipitation in such years. Given that the procedure needs GIS to achieve its best performance, an aerial estimate of the recharge can be obtained, which may subsequently be of use in selecting suitable sites for artificial recharge to be carried out [27].

Chloride balance and estimate of potential recharge in transit

The technique to ascertain the chloride balance and estimate of potential recharge in transit is commonly used in arid and semi-arid regions [28, 29]. The chloride balance in the unsaturated zone, for a long period under environmental conditions whose atmospheric conditions are those of a permanent regime without variations in land use, can be written:

$$P \cdot C_P = R \cdot C_R + U \cdot C_U + \Delta\theta \cdot C_\theta \qquad (8.3)$$

where P, R, U and $\Delta\theta$ are the precipitation, the total potential recharge in transit to the saturated fringe, the surface run-off and the change in soil moisture during the sampling period, respectively, all in mm, and C the mean chloride concentration during the sampling interval for P or the mean concentration for a specific number of samples in the period for R, U and θ as subscripts, in mg/L $- 1 \equiv$ g/m^3.

R is the unknown variable to be derived from atmospheric chloride deposition (AP $= P - C_P$), chloride removed by run-off (AU $= U - C_U$), change in flow mass in unsaturated zone ($\Delta_\theta - C_\theta$), and C_R is measured. U is calculated from data from gauging stations and Δ_θ from daily records of soil moisture measurements in experimental basins. Equation 8.3 can be simplified when the BC is applied to long periods in which changes in Cl in the unsaturated zone and in surface run-off can be ignored. Both conditions are easily achieved in limestone mountainous areas such as the Sierra de Gádor.

The correct estimation of AP, which should include both dry and wet atmospheric deposition, is essential to reduce uncertainties in the estimate of R [30]. Since C_R is the Cl content that moves between the soil and the saturated strip, its value should be measured in very shallow water, hanging aquifers or shallow wells, disregarding anything of atmospheric origin by using chemical and isotopic tracers [31].

The BC method has been applied to the estimation of potential recharge in transit by sampling the hanging aquifers in the area [21]. The abrupt relief in this coastal area and the negative gradient of AP towards the top, between 0.5 and 1 g/m year km, causes the value of C_R to be reduced considerably along the slope. An average recharge of less than 20 mm/year on the coast and about 250 mm/year at the top is achieved. As in the previous case, the data can be manipulated in a GIS environment and the variation quantified.

8.3.5.3 Use of Numerical Models to Estimate Recharge

The ERAS model has been one of the most widely used to estimate recharge [32]. Its principle is the quantification of the actual recharge, R, in a given time interval i by means of the expression:

$$R_i = M \cdot \left(P_i - T_i^\beta \right)^N \cdot A \tag{8.4}$$

where P_i the rain during that period, in mm, T_i is the mean air temperature in °C, and A the permeability of the aquifer in m^2. β is a dimensionless calibration parameter that converts the temperature into ETP. Its value is between 1.3 in cold areas and 1.6 for hot areas. Parameters M and N, which have to be calibrated beforehand using groundwater evolution dynamics data, determine the fraction of the expected effective rainfall that reaches the groundwater level, such as $P_i - T_i^\beta$.

The variation of the water stored in the aquifer is calculated by:

$$\Delta V_i = \Delta h_i \cdot A \cdot S \tag{8.5}$$

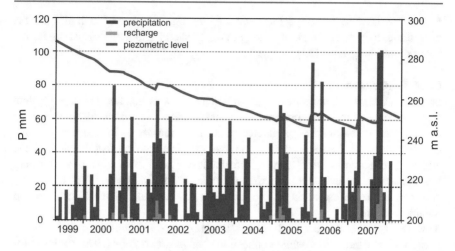

Fig. 8.10 Recharge estimated by the ERAS model in the Ventós aquifer for the 1999–2008 simulation period

where Δh_i is the change measured in the water table (m), and S is the storage coefficient. Since ΔV_i depends on actual recharging R_i (inputs) and pumping B_i (outputs), Δh_i is expressed as:

$$\Delta h_i = \frac{M\left(P_i - T_i^{\beta}\right)^N \cdot A - B_i}{S \cdot A} \tag{8.6}$$

Finally, M and N are adjusted for the simulated period by comparing the values of Δh_i measured with those simulated from the data of P_i, T_i and B_i.

The results from applying the ERAS model to karst aquifers in the provinces of Murcia and Alicante cover a range of between 7 and 208 mm, depending on whether the years are dry or humid, equivalent to 4% and 47% of precipitation, respectively [22]. The variations are due to soil properties, vegetation, geological heterogeneity, and the degree of karstification and fracturing. This same model has been applied [32] to the small carbonate aquifer of Ventós, with low inertia regarding rain events and long periods with a total absence of recharge (Fig. 8.10).

8.3.6 Hydrogeochemistry and Tracers

Hydrogeochemistry and the use of both natural and artificial tracers are exploration and research tools with great potential. The diversity of karst aquifers makes it impossible to derive a simple general rule of universal application without making catastrophic mistakes. The salts that are dissolved in karst waters and their isotopic content tell us about a system's history, preferential flow and a whole series of processes related to hydrodynamics and the lithological nature of its limits. That is why these tracers are regarded as providing useful global information in the medium and long term [33].

8.3.6.1 General
Isotopes are used to determine the age of water or its average duration in the system (radioactivity) or to locate preferential recharge areas. At a more local scale, they can be used to delimit the perimeter of a protection zone. In addition to ionic concentration and their spatial and temporal variation, conical features and, above all, deltas can reveal useful genetic information. This allows areas of contrasting hydrogeochemical behaviour, mixing processes, ionic exchanges and even residence times to be established. This involves the determination of accurate and reliable analyses.

Karst exploration traditionally used artificial tracers, although its main objective was often to determine possible connections between two points. The potential for the use of tracers in karst is very great and constitutes an indispensable tool in studies on the propagation of pollutants [4, 5] in Chap. 1 [28]. In addition to the traditional colour and chemical tracers (CI, Na), lithium is very conclusive, as it has such low concentrations in the natural environment [34].

8.3.6.2 Box 3: Examples of the Use of Tracers

Nastan-Trigrad karst system, southwest Bulgaria
The Nastan-Tigrad karst system has already been described in Chap. 5. It is a region in the Rhodopes, in the southwest of Bulgaria near the Greek border. It is characterized by an abrupt karst relief. The system discharges from several springs with flows that can exceed 2 m^3/s (Fig. 8.11 and Table 8.1), two of them (Springs 67 and 39) under quality control and flow by the Bulgarian public body (NIMH). The objectives of this experiment were to confirm the results of previous experiments, in the sense that the River Tenesdere feeds the Beden spring (Spring 39), as well as contributing to knowledge of the geometry and compartmentation of the system.

Table 8.1 Main characteristics of the sampling points. M: spring; S: borehole

Parameter	M 10	S 101	M 67	S 11	M 39a	S 12	Tenesdere river
Flow rate l/s	20	–	582	97	720	1	721
Dimension m	760	770	709,9	780	785	820	1260
Distance injection point (km)	8.75	8.8	9.25	7.25	7.0	10.75	0
Difference in height, m	500	490	550	480	475	440	0

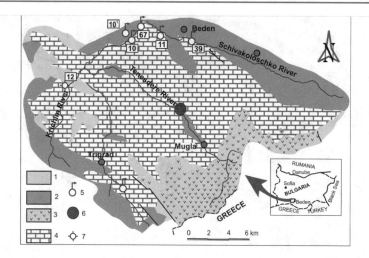

Fig. 8.11 Hydrogeological scheme of the Nastan-Trigrad karst system: **1** gaps, conglomerates, sandstones and limolites; **2** granites, gneisses and schists; **3** ryolites; **4** marble; **5** spring; **6** injection point; **7** drilling

Two previous experiments were carried out in this sector in August 1954, with NaCl as the tracer, and in July 1999. In the first one, tonnes of common salt were introduced close to the site used in this experiment. The tracer was detected in Springs 10 and 67 after 41 and 63 h, respectively.

On 9 June 2000 at 12:00, 5 kg of diluted fluorescein were injected into the riverbed (Photos 8.1a and b) in its upper basin, 200–300 m above the bed, 3.5 km from the village of Mugla, at an altitude of 1260 m (Fig. 8.11). The river flow at the time of injection was 721 l/s (Table 8.1). The output of the tracer was continuously controlled at Points 11 and 67 in the fish farms at Beden and Nastan (Photos 8.2 and 8.3) by means of automatic samplers (Edmund Büler, model Calipso 4.20.2–04). Sampling was conducted on an hourly basis. Items 10, 12 and 39a were sampled every 12 h and bags of activated carbon were placed in each. In addition, one litre of zero-concentration water was taken to estimate the background at each of the injection points. The water samples were analysed by means of a spectrofluorometer and a spectrophotometer (SPEKOL 10, Carl Zeiss Jena with a determination limit of 5 ppb). The fluorescein from the activated carbon samples was diluted using 100 ml of a 15% KOH solution for 15 min. The characteristics of the sampled points are shown in Table 8.1.

Spring 67 is located in a fish farm on the Nastan-Devin road. The sub-caudal pressure, assumed to be constant during the experiment, was 582 l/s. Ninety-two water samples were taken between 10 and 14 June, and 11 further samples, at the rate of two per day, between 14 and 19 June. Three bags of activated carbon were also used, and the last was removed on 1 July, 23 days after injection.

Photo 8.1 a,b. Two snapshots of the discharge of the tracer into the river Tenesdere

Photo 8.2 Drs Vallejos and Calaforra, together with a Bulgarian colleague, prepare the active carbon bags (Photo A. Pulido)

Photo 8.3 Tracer exit at the fish-farm spring (Photo A. Pulido)

Spring 11 is also in a spring that supplies a fish farm. This, too, was equipped with an automatic sampler. With a flow rate of 97 l/s, 75 water samples were taken between 10 and 14 June. Also 11 samples were taken at the rate of two per day between 14 and 19 June and activated carbon was used until 1 July. At Spring 39 at Beden, 19 samples were taken at the rate of two a day between 10 and 19 June, as well as three bags of activated carbon. Its average flow during the experience was 720 l/s.

Another 19 samples were taken from Termal (Nos 10 and 110) over the same interval, nine from Borehole 101 and 10 from the spring on the left bank of the River Vacha. These had three bags of activated carbon in the boreholes and another at the spring. Six samples were taken from Borehole No. 12 between 11 and 14 June, and one bag of activated carbon was used in that period.

The values obtained were represented in the respective restitution graphs and subsequently interpreted using the Traci program, version 4.0.4β, which enables the user to adjust 1D models that give information on flow velocities and the dispersion in the direction of flow.

The restitution graphs of Springs 11 and 67 are shown in Fig. 8.12; in both cases an adequate reconstruction of the tracer's transit is achieved. They are the same springs as studied previously. Fluorescein was seen with the naked eye at 13:45 on 11/06/2000 at Spring 11, while in Nastan (Spring 67) it was detected analytically at 9:30 on 11/06/2000.

At Spring 67 the maximum value–32 ppb–was reached at 18:30 on 11/06/2000; that is, 54.30 h after injection. The apparent speed in this case is 170 m/h. In Spring 11 the highest value was detected around 21:30 on 11/06/2000, which gives an apparent average speed of 126 m/h. As it can be

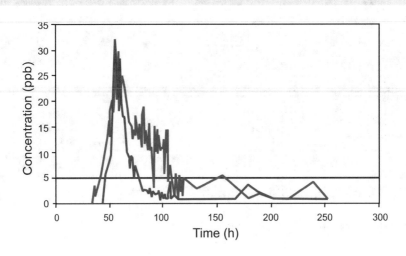

Fig. 8.12 Restitution graphs at Spring 67 (blue) and Spring 11 (red)

seen in Fig. 8.12, the average speed in the direction of Spring 11 is lower than in Spring 67, although detected earlier because it is closer to the injection point. In Springs 39a, 10 and 12, the observed values were below the detection limit during the entire experiment (Fig. 8.13) and therefore the arrival of the tracer at these sites cannot be guaranteed. Everything indicates that the upwelling at Outcrop 39 is disconnected from the injected system and, consequently, its feeding area is different from that at Springs 11 and 67.

The application of the above model to Springs 11 and 67 (Fig. 8.14) provided information on the hydraulic characteristics of the system. The transit speed is higher for Nastan (Spring 67) with a sharper curve and lower dispersion value; almost 30% of the tracker was recovered (26%, in fact). Between the two, it was possible to recover almost 40%, which together with the high measured speeds indicates a highly karstified system with preferential flows. The fluorescein concentrations measured in the active carbon bags corroborate the results from Springs 67 and 11, and there seems to be

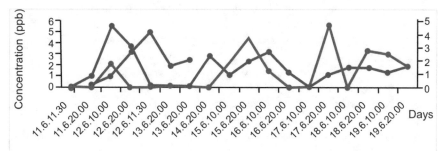

Fig. 8.13 Possible restitution graphs of the track marker in Outcrops 39a (blue), 12 (green) and 10 (red)

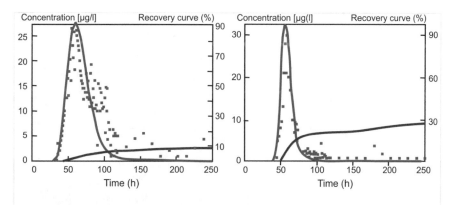

Fig. 8.14 Simulation of the restitution graphs for Springs 67 and 11

some connection with Spring 10, adjacent to them. In the experiment, the average transit time for Nastan was 56 h, with an average transit speed of 167 m/h, a dispersion of 0.005 and a tracer recovery of 26%. For Beden, the average transit time was very similar (56.5 h), but the transit speed was much higher (128 m/h), as well as its dispersion (0.22). By contrast, the recovery was notably lower (10%) [35].

As a curiosity, an experiment can show how it is possible to have several exits from a single injection point, although the fact that it is a river means that the possibility of infiltration in areas of unequal influence increases significantly. Previous studies of the flows of these occurrences by correlation and spectral analysis have concluded that slow flow largely prevails over fast flow, with a memory effect estimated at 64–83 days, in apparent contradiction with what is deduced here. The explanation is the notable difference between the behaviour of a karst system as a whole and its behaviour in detail, as in the case of the possible upwelling river relationship. This points to the need always to carry out multiple approximations in karst hydrogeology studies to limit the necessary simplifications involved by each particular approach.

Finally, it should be noted that the apparently unrecovered tracer fraction may be due to a combination of several processes; first, the long tail of the restitution curve, due to the effect of smaller existing discontinuities; the possible presence of other unsampled outlets; and the possible entrapment of a fraction of the contaminant, as described in several cases.

Dobrich, northeast Bulgaria

An experiment at Dobrich was carried out on the Sarmatian Aquifer or Upper Aquifer [36] in northeast Bulgaria (Fig. 8.15). Most of this area is occupied by calcarenite materials, limestones and detritic materials, including Miocene loess. The impermeable substrate consists of clays and loams from the Hauterivian [37].

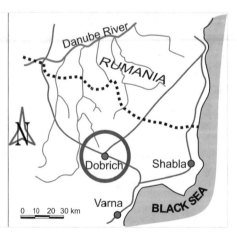

Fig. 8.15 Location of the Dobrich sector

Photo 8.4 Preparing the injection of the tracer into the borehole at Dobrich Park. On the left, Dr Dimitrov, and on the right Dr Velikov (Photo A. Pulido)

The experiment consisted of injecting a tracer into two shafts (Photo 8.4) of known lithology and which had already been studied [36] in the summer of 1995. The first, located south of the city of Dobrich near the bus station, is 118 m deep with a static level of 63.73 m deep. The second is in an urban area and is 47 m deep and with a level of 19.85 m deep (Figs. 8.16 and 8.17).

Prior to the tracer procedure, resistivity and natural gamma emissions were recorded using a pair of channel scintillometers, one lithological and the other volumetric, calibrated in CPM and microR/h respectively. Readings were recorded of the calimetry, temperature and equipotential. The layout of the column was made with 131-iodine (dissolved) (Photo 8.5) and 37 MBq of the activity mixed uniformly throughout the water column, and was recorded with a two-channel gamma spectroscope. Several active gamma radiation profiles were carried out in the water column using the high-sensitivity lithological channel, with the volume channel used as a control. The isotope, mixed with non-radioactive NaI crystals, was introduced by syringe into the injector, previously filled with NaCl as a buffer. Gamma-gamma radiation profiles were measured with a two-channel scintillometer using Americium 124 as a low-energy (60 keV) radiation source with an activity of 4.10^9 Bq to provide information on the rock's density.

The records obtained are shown in Figs. 8.16 and 8.17. By tracing the column by the single-well method [38, 39] and after logging at various times

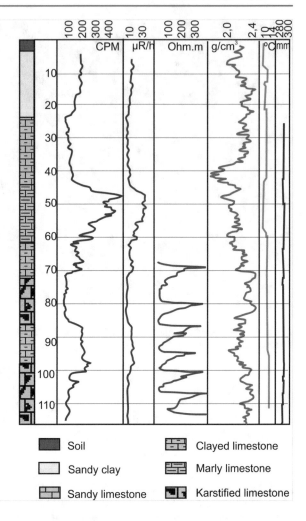

Fig. 8.16 Lithological column of the survey at the bus station

Soil	Clayed limestone
Sandy clay	Marly limestone
Sandy limestone	Karstified limestone

t_i can determine the specific flow q (*m/d*) or filtration velocity in both wells by the expression:

$$q = \frac{1,81d}{m(t_2 - t_1)} \log \frac{C_1 - C_0}{C_2 - C_0} \qquad (8.7)$$

where d is the borehole diameter (m), t_i is the time at moment i, C_i is the gamma activity (CPM) at time t_i and C_0 the bottom activity, and m is a coefficient that is obtained from the velocity of the water in the well and the velocity in the formation. The values obtained in the park borehole, in the 29.5–44.5 m depth range, were between 0.26 and 12.38 with an average value of 6.12 m/d. The values obtained in the park borehole, in the 29.5–

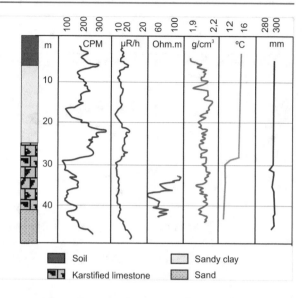

Fig. 8.17 Lithological column of the park borehole and records of resistivity, calimetry and gamma rays

Soil

Karstified limestone

Sandy clay

Sand

Photo 8.5 Container with the tracer (Photo A. Pulido)

44.5 m depth range, were between 0.26 and 12.38 with an average value of 6.12 m/d. In the borehole of the bus station, values of 0.0091 and 2.04 were obtained, with an average value of 0.472 for the Sect. 63.5 to 114.5 (Fig. 8.18). For the same depth intervals, the density values obtained varied between 1.995 and 2.125, with an average of 2.062 g/cm^3 at the park, and between 2.120 and 2.440 with 2.278 g/cm^3 at the bus station (Fig. 8.19).

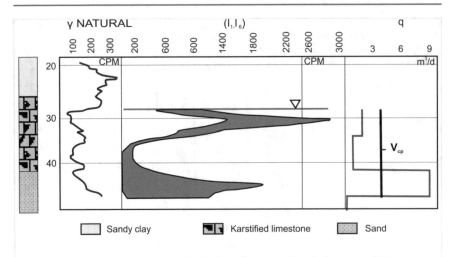

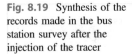

Fig. 8.18 Synthesis of the records made in the park survey after the injection of the tracer

Fig. 8.19 Synthesis of the records made in the bus station survey after the injection of the tracer

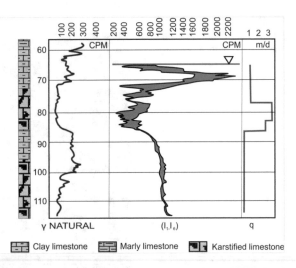

The values applied for C_0 ranged from 84 to 190 with an average value of 127.7 CPM for the park borehole, and 20 and 210 with an average value of 95.7 at the bus station. The most transmissive tranches in both boreholes were tranches 36.5–44.5 m, with an average q of 10.51 m/d, ρ_a of 2.05 g/cm^3 and C_0 of 120.7 CPM. For the bus station, the Sect. 63.5 to 84.5 m, with an average of 1.01 m/d, ρ_a of 2.29 g/cm^3 and C_0 of 68.31 CPM. The actual flow velocity U can also be estimated from q by means of $U = q/n_0$, where $q = V = K \cdot i$ where K is the hydraulic conductivity in m/d, i the hydraulic gradient and n_0 the effective porosity. The values obtained were 14.4 m/d for the depth range of 29.5–44.5 m in the park borehole.[2]

8.3.7 Simulation Models

Although they cannot be considered as an exploration tool, simulation models contribute to a policy for the rational exploitation of a karst aquifer's resources. This is because they permit all the possible responses, factors and total pumpable volume, and their effect on a spring in Chap. 4 [16], to be determined, as described in Chap. 5.

The methods capable of being applied are highly diverse. Conventional versions, applied to a porous medium under the concept of Representative Elemental Volume (VER) and using the technique of finite differences, can give very good results if the aquifer is of the Torcal type. This same idea, superimposed on the conceptual model of a karst aquifer of blocks and ducts with hierarchization (the 'intermediate' model), has given good results in the simulation of Mediterranean karst aquifers. More complex and more precise are the models involving bi- or three-dimensional finite elements that theoretically allow the simulation of an entire network of karst ducts, blocks and permeability [16, 40]. The black-box or 'grey-box' models can provide a suitable approximation, and some can simulate pumping from the flow and predict the response.

8.4 Exploitation Methods

8.4.1 Background

In the karst environment there is great heterogeneity and anisotropy, and a great variation in flow over time. From this it can be deduced that there is regulation that optimizes the flow, on the one hand, and on the other that heterogeneity plays an important role. Both can lead to a high percentage of exploitation failure, and there greater risk than in detritic aquifers. Regarding this regulation, we will look first at the simplest types of water extraction (Fig. 8.20). These are diverting a spring, sometimes building a reservoir at its source, excavating a gallery, which was the most widespread procedure for many centuries, and constructing pools. They can be combined, creatively or otherwise.

In this diagram, we now move to the highest risk. The regulation of surface water in the karst medium is not the object of this study. However, it is in this environment that work such as flood abatement is carried out and fulfils its purpose. Regulation and storage, however, can pose serious problems since it is difficult to construct a watertight tank. This is the case of the Montejaque reservoir in the province of Malaga, and the Isbert reservoir in the province of Alicante, to name but two.

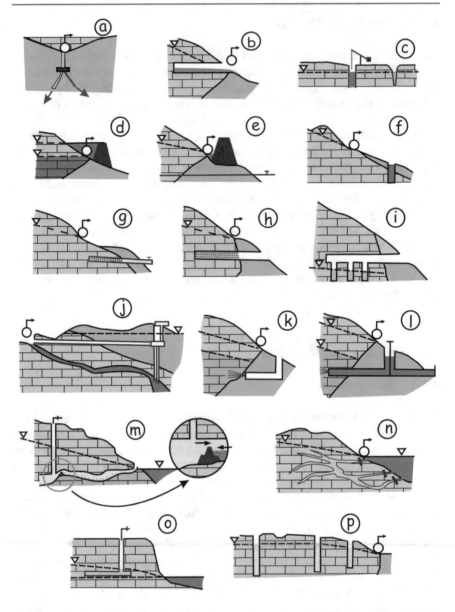

Fig. 8.20 Main methods of water capture, in approximate chronological order: **a** diversion of a spring; **b** gallery excavated into a spring; **c** extraction from well; **d** reservoir in spring; **e** complemented with horizontal drills with capping under the dam; **f** vertical borehole downstream of the spring; **g** horizontal shaft with capping; **h** gallery downstream of the spring; **i** gallery with boreholes inside (Suizos, Crevillente); **j** vertical borehole and horizontal gallery (Lez, Montpellier); **k** borehole and gallery with holes under the spring; **l** As in k, with horizontal elongation and capping; **m** underground dam and upstream borehole (Port-Miou); **n** waterproofing of submarine karst ducts; **o** vertical borehole and horizontal gallery at sea level (coastal aquifers); **p** boreholes upstream of the spring

8.4.2 Diversion of Springs

Many karst springs pose a serious problem at the point where they are made to divert, for instance the source of the River Mundo (Photos 8.6a, b and 8.7) or the Gorgotón in the bed of the Júcar (Photo 8.8). The system simply consists of diverting the spring's flow, and it is recommended to install some form of perimeter protection. It is the optimum solution where the demand for water varies roughly with the flow of the spring, because it does not involve much expense. This would be the case of the Chorrador spring in Alcoy (Photos 8.9 and 8.10), which yields a flow of between 20 and 70 l/s, previously supplying the city, which at that time had a noticeably constant demand of 300 l/s.

It is necessary to clean the area, to locate the precise point or points of emergence and to arrange canalization, and for this reason it is convenient first to install a limnometric scale. In the case of collection for an urban water supply, it is strongly recommended to erect a hut and construct a closed conduit to avoid pollution during transport. However, what happens in most cases is that water demand remains constant over time, or varies in an inverse way to the spring, so it is high when flows are minimal (summer) and minimal in peak flow. This is why

(a)   (b)

Photo 8.6 a,b. Two views of the River Mundo's source, in a unique 'blowout' (Photos A. Pulido)

Photo 8.7 Source of the River Mundo, showing high flows of short duration: an example of a spring with little prospect of regulation, given its characteristics

Photo 8.8 Gorgotón spring, within the bed of River Júcar (Photo A. Pulido)

Photo 8.9 Molinar spring, traditional water supply of the city of Alcoy (Photo A. Pulido)

Photo 8.10 Alcoy water-supply shaft well, in the vicinity of Molinar spring, to meet demand at times of pronounced low-water levels (Photo A. Pulido)

regulation has to be imposed, and to do so we need to know the characteristics of the underground reservoir, its operation and the evolution and amount of demand. We must not overlook the intended environmental use of the spring water when planning its exploitation.

8.4.3 Reservoirs Formed Within Springs

There are many examples around the world of reservoirs having been built within a spring (Photos 8.11a and b, near Lisbon), and those in former Yugoslavia date back to the time of the Dioclesian Emperor. They consist of a real reservoir constructed right at the outlet of the spring, which manages to retain water and, more importantly, to increase the volume that is dammed to a certain extent. The aquifer itself acts as the reservoir. The arrangement has the following advantages:

- there is no danger of catastrophic flooding in the event of accidental rupture of the dam
- there is no risk of clogging the tank
- the evaporation losses are negligible
- geotechnical problems are negligible, as the requirements for strength and stability are minimal
- low cost.

Before carrying out work of this type, it is necessary to have a good knowledge of the operation of the system and its parameters, and to be certain that when the level is increased there is no exit through another point at a lower level beyond the downstream boundary. The fact that the average storage coefficient is of the order of 0.01–0.05 is also regarded as a disadvantage.

8.4.4 Galleries

Galleries were the preferred means of water capture before lifting equipment was developed. As a regulating element of aquifers, a gallery is notorious for being an expensive solution, except in very special conditions, due to the fact that the aquifer's regime is not modified. This is to say that more water continues to flow when it is less needed, and the flow is less in times of low water, while the volume of the store is reduced. However, they have the apparent advantage that, projected perpendicularly to the set of fractures with the most hydraulic activity, they can intersect a greater number of water discontinuities than any vertical work.

Galleries' use can be justified if they can be made of impermeable material, and if there are sluice gates that can be operated at will. If this is not the case, what is created is a *bleed* in the aquifer, manifested by a gradual decrease in the flow or even totally drying up. This is the case of the Galería de los Suizos (Photos 8.12a and b) in the Crevillente mountain range, 2360 m long and some 3 m in diameter,

Photo 8.11 a,b. Two 'reservoirs' in karst springs near Lisbon; the second used to supply the local paper industry (Photos A. Pulido)

(a)

(b)

Photo 8.12 a,b. Access to the Galería de los Suizos on the southern edge of the Sierra de Crevillente (Photos A. Pulido)

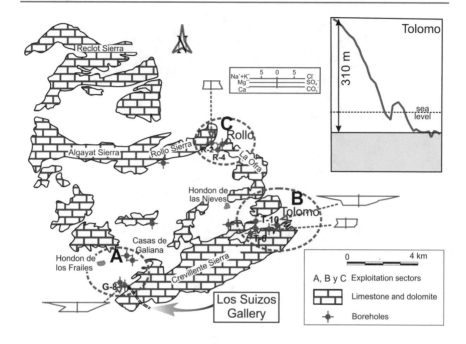

Fig. 8.21 Hydrogeological scheme of Crevillente mountains, showing the Galería de los Suizos and Stiff diagrams

which, after crossing some loams, goes through the limestones that make up this mountain range. Recently constructed, it first had a flow of more than 500 l/s. After a short time, it worked only during the rainy season. At present (Figs. 8.21 and 8.22), there are 12 boreholes inside the tunnel that extract a flow of about 400 l/s throughout the year, with the piezometric level at more than 100 m deep.

8.4.5 Boreholes

8.4.5.1 Practical Considerations

The most usual form of regulation consists of mechanically sunk shafts, with varied characteristics, operated to meet an intended objective. As a usual rule in carbonate materials, it is recommended that the work is carried out by rotary percussion (Photo 8.13) or percussion, although this is very slow. Reverse rotation and normal rotation can also be used, but with water. The method of rotation with normal sludge (bentonite) must be abandoned, because then a borehole is difficult to sink and is very expensive if a loss develops.

The practical considerations in sinking a borehole include the following:

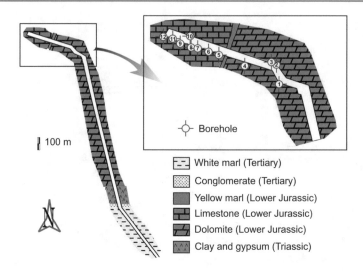

Fig. 8.22 Gallery on the ground plan, showing the land crossed and details of the boreholes currently in its interior

Photo 8.13 Rotary-percussion machine sinking a borehole in the marble of Castell de Ferro (Granada) (Photo A. Pulido)

- To protect the lifting equipment, the pumping chamber must always be tubed. The depth of the pumping chamber can be estimated from the depth of the water plus the magnitude of the annual and interannual fluctuation plus the depression due to pumping plus a margin for extreme years and eventual overexploitation. It is essential to know to a reasonable order of magnitude each of these data when scheduling the work. The rest of the shaft will be left untubed, in most cases.
- In *percussion drilling* the resulting detritus is analysed, in what is known as a shaft-driller's way of working, to get an idea of the most productive sections. In general, the more the sample appears to have been washed and the thicker its edges, the greater its productivity. If is not possible to extract detritus to a depth of several metres, this is generally a guarantee that the work will be highly productive.
- It is necessary to carry out valving tests starting from no pressure, and their results will give an idea of the changes in the productivity of a shaft. It is of no use to carry out tests once it has been verified that the level does not go down or that it even goes up. This would be the case when the borehole is cleaned by the extraction.
- When programming a borehole, at least 100 m of saturated zone must be pre-viewed, as there will always be time to limit this afterwards. Valves and/or production tests of any other type will advise on reduction or continuation.
- When working for official bodies (or even individuals), on the preliminary draft of the work it is advisable to recommend all the operations that can potentially arise, such as re-drilling, constructing foundations, increasing the diameter and performing acidification. This is because these organizations usually work to an approved budget that is later very difficult to modify.

Before selecting the drilling site, it is necessary to have as precise an idea as possible of the operation and characteristics of the system's reservoir (situation and level of springs, depth of the piezometric level, performance of nearby catchments, hydrogeochemistry, etc.). For precise siting, electric geophysical prospecting in SEV and trial pits can help considerably. Especially if there is a mud coating, these enable the most fractured areas to be detected, manifested as a fall in the apparent resistivity values.

Consultation of a previous photogeological borehole is essential to determine the most fractured areas, verifying the findings by observations on the ground. Bear in mind that it is normally best to tap the water at the greatest possible depth, so investigate low areas. Ravines meet these conditions best, yet they tend to correspond to weak areas, as they are normally fractured. If there are any doubts about the continuity beneath a mud coating, always start drilling in carbonate material, because this will avoid any surprises. Although it involves extending a path and perhaps resorting to drilling further holes, it is usually cheaper than sinking a 'negative' borehole.

What flow can be obtained? It can be very variable from one point to another, from the *blocks* to the *ducts*, with no apparent relationship between the depth of the work and the thickness of water-bearing rock captured. However, in well-programmed works it is relatively easy to obtain medium to high flows (30–50 l/s), with relatively high specific flows (5–10 l/s/m). Unanticipated results are frequent, due to both excess and deficit: flows of less than 5 l/s and greater than 100 l/s. Next to the Spring of the Saints, the SGOP sank a shaft that supplied 350 l/s with 14 cm of depression. Saturated travertines are among those capable of providing greater flow, with greater spatial constancy.

8.4.5.2 Development by Acidification

General Aspects
Acidification is the optimal procedure for the stimulation and development of a carbonate aquifer (limestones and dolomites and sandstones and conglomerates of carbonate cement). The most commonly used acid is HCl, since SO_4H_2, although cheaper, yields SO_4Ca as a by-product, elimination of which may pose problems. The simplified reactions are:

$$CaCO_3 + 2HCl \rightarrow CaCl_2 + CO_2 \uparrow + H_2O \quad \text{and}$$
$$MgCO_3 + 2HCl \rightarrow MgCl_2 + CO_2 \uparrow + H_2O$$

It is of most interest in the case of limestones where parts are fractured. Otherwise, the only result that is achieved is to clean the environment and minimally increase the radius of the work. Bear in mind that one litre of 15% HCl (commercial-grade HCl has 20–20° Beaumé → 31–35%, so needs to be diluted, always introducing the HCl to the water and not putting the water into the HCl, to avoid accident). It is convenient to use a base. Sodium bicarbonate in water (or in powder, to prevent accidents) dissolves 221 grams of $CaCO_3$ and 203 g of $MgCO_3$. At 25° C, the reaction with limestone lasts 40 min, and with dolomite 50 min (at 95% acid neutralization). It is usual to work with several tonnes (10–20 t), carrying out several tests, generally with increasing doses.

Several additives may be added to HCl, for the purposes below:

Reaction retarders—To ensure that the reaction occurs only once it has been introduced into the aquifer. $CaCl_2$ is usually used in variable proportions (0.5–7%, 150 kg in 5000 kg), as estimated, or amyl alcohol (g/l). If a first small cleansing insertion is made, there is no need to add any retarder as the $CaCl_2$ generated will serve that purpose.

Iron and aluminium oxide stabilizers—To produce gelatinous hydroxides that can block fissures, these are put into solution by an acid and then neutralized. Citric or lactic acid is used in concentrations of 2–10 g/l to stabilize the Fe and Al that are normally present in limestone.

Anti-foaming agents—CO_2 can create large amounts of foam, which can cause acids to escape from the borehole exit. Usually the agent is used at the rate of 1–10 g/l of amyl alcohol. It is sold under various trade names.

Inhibitors—These reduce the aggressive effects of acids on pipes, pumps, and so on. Phosphates, polyphosphates, thiophenols, polyamines, gelatine, etc. are used at a concentration of 2–5 g/l. Since they degrade readily, they must be added to the mixture immediately before use.

Since gypsum is usually present alongside limestone (but not always), it is necessary to add a compound (ammonium bifluoride, NH_4HF_2), which gives rise to ammonium sulphate, soluble $SO_4(NH_4)_2$, to prevent it from precipitating. There are some other additives that can be used, but they are not really essential.

The effect of acid (20,000 kg of 15% HCl dissolves 4420 kg of $CaCO_3$, 1728 m^3) is due more to the opening up of fissures and cleaning them than to any increase in the diameter of the shaft, the effect of which would be negligible. That is why a key element of acidification, to be taken advantage of, is the great pressure created by the release of CO_2, which can cause actual hydraulic fracture.

Carrying Out Acidification

Ideally, acidification is carried out under pressure in the shortest possible time (Photos 8.14 and 8.15), with elimination of the acid immediately after neutralization or a little before. The work should be hermetically sealed to facilitate both the introduction of the acid into the aquifer and the action of CO_2. Several doses are administered, the first normally cleaning the hole, using its volume, and the following two a rather greater extent. More than three doses does not produce any

Photo 8.14 Before acidification, it is necessary to extract the pumping equipment; the photo corresponds to a shaft in Aspe (Alicante) sunk by means of explosives (Tolomó sector) (Photo A. Pulido)

Photo 8.15 View of the
wellhead with the system of
acidification (Photo A.
Pulido)

significant improvement (exponential decrease). Two orthodox procedures are usually used:

1. Pressurized injection with hermetic closing of the borehole mouth, so that CO_2 creates the pressure
2. Injection without a hermetic seal, with the introduction of a simultaneous flow of water to ensure that the acid penetrates the aquifer.

(a) *Hermetically sealed injection*

   The borehole must be hermetically sealed and the pipe cemented in place to avoid unanticipated results. In doing so, it is necessary to use: a manometer to read the pressure in the working; a connection to introduce the acid from above or, what is more usual, a tube with a basal exit or lateral opening; and a connection for water, either sprayed or hosed, that has valves to open or to close it shut. It is vital for the injection pipe to have a transparent section so the direction of the flow within can be seen, to detect if at any time there is a rise from the borehole towards the tank.

   The injection of acid can be achieved by either canisters or gravity. Often, the acid is delivered from a truck (Photos 8.16 and 8.17), so it is necessary to do it quickly to avoid the CO_2 passing into its tank and the acid overflowing and escaping. Once the acid has been put in and all the valves have been closed, it is introduced into the aquifer using a compressor. If the water level is close to the

Photo 8.16 Acidification of a borehole in Sierra del Rollo, Aspe (Photo A. Pulido)

surface, it is essential to take precautions to avoid any excessive pressure that could cause the pipe to jerk. To immobilize the section to be acidified, two packers can isolate the sector that is being worked on, to improve the yield of the borehole.

With the help of inhibitors, the acidification can be undertaken with the pump in place. After approximately an hour of injection, all the Cl_2Ca generated can be pumped out and eliminated; if near the surface, a compressor can be used.

(b) *Injection without hermetic seal*

Without a hermetic seal, injection is less effective, because it cannot take advantage of CO_2 pressure. The procedure consists of putting in the acid, either by gravity or with a pump, and making it penetrate by introducing a continuous flow of water (3–6 l/s). This procedure is applicable only at great depth (more than 80 m down), otherwise there is a risk of acid flooding back out through the borehole exit (Fig. 8.23).

Acidification is usually quite laborious, and only few companies in the market have the equipment necessary to undertake the procedure. Theoretically, it just

Photo 8.17   Other view of the same borehole in Sierra del Rollo, Aspe (Photo A. Pulido)

Fig. 8.23   Highly simplified
diagram of acidification with
a hermetic seal

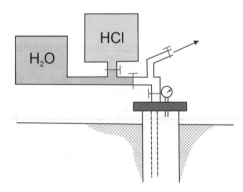

needs pumps, compressors, tanks of acid and of water, pipes, sprayers, and so on. In
reality, even if we can do without any additives, it involves two trucks, one for acid
and one for water, their respective tubes, the need to seal off the borehole mouth, a
manometer and sets of closing valves.

Increasing Performance by Acidification

The increases in flow obtained by the acidification process are usually spectacular. They are rarely less than 100 and 500% or more is common. The best is in specific flow: a specific flow in Sierra Grossa of 0.1 l/s/m increased to 0.4 l/s/m after introducing 13,000 kg of ClH, and from 4.8 to 5.6 l/s/m in another case. It is almost always the case that the greatest increases are in wells that had a very low initial flow. According to Koenig's studies of a large number of acidification procedures, 80% achieved significant improvements in specific flow (0–1100%), but there was no improvements in the other wells, or even a counter effect.

Tests carried out in 15 shafts by another study show that the most spectacular improvements correspond to works with the lowest starting Qs. The most notable increases are achieved after the first acidification, with many shafts reporting a decrease after a third procedure. The improvement overall was between 17 and 1100%. Similar was found by studies of 12 wells, where the most spectacular results corresponded to the second procedure (15–1875%).

It is important to point out that the great CO_2 release confers a high level of aggression on the water. Together with the strong CO_2 release, this means that subsequent water analyses (even more than a month later) show water with a high HCO_3 content—and calcium chloride facies. On the other hand, since acid has a greater density (1.21 g/cm^3) than water, it tends to sink to the bottom. For this reason, the best injection system has lateral jets and a closed bottom.

8.4.6 Other Exploitation Systems

Other ways to extract water include that the case of the Ras el-Ain springs, 5 km from Tyre and 700 m from the sea. They are fed by the limestones of the Cenomanian–Turonian at much greater altitude and well up in marl along a fracture, so they are of artisan origin. The Phoenicians built water towers in which this level rose to 6–8 m, allowing the supply to be distributed across a wide coastal strip (800 l/s). In Greece, also, there are interesting examples, such as Kiveri (Photos 8.18 and 8.19), which has the problem of mixing with seawater.

Another great work of engineering, although it is not clear if it is justified, is the prolonged investigation into tapping the submarine springs of Port-Miou (Cassis, near Marseille; Photo 8.20) [2]. To reach the catchment has involved many hours of diving to map the duct network, involving geophysical methods for temperature control, speed, conductivity and multidisciplinary equipment. It consists of an underground dam in a bell, to prevent the entry of sea water, the removal of residual salt water and the extraction of the fresh water by means of a well. The system never came into operation since it did not achieve 'fresh' water on a reliable basis. Attempts have also been made to tap the upwelling water by installing underwater bells with conduits leading to the surface, from where it can be stored and possibly piped ashore.

An example similar to Port-Miou is Golubinka springs [41], used as an urban water supply. The springs are close to the sea, which at times of low water greatly

Photo 8.18 Spring captured at Kiveri (Greece) (Photo A. Pulido)

Photo 8.19 A detail of the spring showed on Photo 8.18 (Photo A. Pulido)

Photo 8.20 Submarine spring of Port-Miou (Photo A. Pulido)

increases the water's Cl content. Following hydrogeological and morphological studies of the reservoir and an underwater topographical survey of the gallery linked the spring to the sea, an underground dam with regulating gates was built, which led to a significant reduction in the Cl$^-$ content; however, the flow was reduced.

Experiments with tracers showed that there were two further major conduits feeding the upwelling. Two new dams made it possible to raise the water level at the source, reduce the losses to the sea and keep the Cl$^-$ content below 250 mg/l. Through this project, the saline content of the pumped water was significantly reduced. However, this scheme has never been operational because of the insufficient estimates of the increased demand for water.

Attempts to capture coastal freshwater have also been made by closing ducts that connect to the sea to ensure good water quality (as in the case of the Golubinka springs [41]; Fig. 8.20n), or by combining wells just above sea level [42, 43] in order to avoid salinization (Fig. 8.20o). In these cases, the saline concentration of the water is often higher than that of the aquifer, even at +2 m, indicative of continental water from above sea level somehow mixing with sea water.

Another means of exploiting underwater springs is the experimental extraction at the Welsh spring in the Tyrrhenian Sea (Italy). The mechanism consists of placing a bell over the point where the water wells up, already pinpointed by other methods, in such a way as to ensure that the fresh water flows directly out through a tube positioned at its upper end, preventing salt water contamination. Under these conditions, the water rises to a higher level than that of the sea, making it easy to tap.

8.5   Impacts of Exploitation

Two main types of impact of exploitation are recognized: **direct** impact; that is, strictly linked to the action of exploitation; and **indirect impact,** depending on a series of circumstances. Frequently, these impacts are associated more with so-called *overexploitation* [44, 45] or, as some prefer, *intensive exploitation* [46–48], although this is not necessarily the case.

8.5.1   Direct Impacts

Descent of piezometric levels—This is the immediate effect of any extraction of water, and it affects the environment of the borehole in question and spreads to wider sectors depending on the characteristics of the formation exploited and the total quantity extracted. This descent may be occasional, as in sporadic pumping, or almost continuous, although affected by the pumping and the system's feeding regime. Continued extraction undertaken well in excess of the system inputs may lead to mining of the aquifer, almost exhausting the system [49] in Chap. 1 [19].

Mine drainage leads to a decrease in piezometric levels, with a consequent effect on the environment. With proper planning, the risk can be quantified. A clear example is the case described by [50] in Hardee County (Florida) of phosphate mining in a karst area, in terms of the performance of the boreholes as urban supply, for agriculture and for the chemical industry there.

Compartmentation of the aquifer—In areas of complex tectonics with an impermeable substratum of very irregular geometry, exploitation can generate descents that reduce the piezometric level below certain impermeable thresholds. This gives rise to compartmentation, whereby subdivisions are formed in the original system. Some sectors have hardly any descents and others see a notable increase due to the effect of barriers of an impermeable nature have on these descents (negative barriers, with double descents). A classic example is the Quibas System in the provinces of Murcia and Alicante [51], initially considered to be a single system. After intensive exploitation, up to seven subunits of well-differentiated behaviour have been identified, both in their piezometric evolution and in the quality of their water.

Increase in operating costs—Any decrease in levels leads to an increase in the depth required in a shaft, which in turn is related to the energy requirements. There are cases in which the required depth increased from 50 to 250 m in less than a decade, trebling the cost of extraction. This is the case in the Crevillente aquifer, already mentioned.

Deterioration of water quality—Sometimes the exploitation of an aquifer mobilizes water of poor quality, for example in the outermost sections of the system, due to gravitational segregation. This is often so in the numerous karst systems in the Betic Cordilleras that have the Keuper Trias as a substratum. In areas with vertical hydrogeochemical zoning, after a few years of exploitation there are instances of a transition from calcium bicarbonate to sodium chloride facies.

Much more frequent is the exploitation of a coastal aquifer promoting the advance of a front of saline water into its karst system. This is the case in the Campo de Dalías aquifer (Almeria, Spain), where continuous pumping has created sectors with more than 10,000 µS/cm [52]. *Seasonal overpumping* with marine intrusion induction can sometimes occur when the affected coastal aquifer is small, although the extraction is of less volume than its annual recharge [53].

Deterioration related to the drainage of mining areas on carbonates is a special case. The Olkusz mines in the vicinity of Krakow (Poland) provide a good example —indirect, to some extent—of pollution from the waste from a paper mill more than 5 km from the mine [54]. The mobilization of fine particles—clays and silts—by the turbulence generated by pumps is another aspect to be taken into account, making it essential to decant or filter the water. This makes it advisable to control the turbidity during any testing to evaluate the aquifer potential of a sector, especially if it involves pumping from the spring itself or from a karst duct.

Abandoned boreholes—Intensive exploitation can lead to boreholes being abandoned due to problems of either water quality or quantity. This is the case in many coastal aquifers, as already indicated. In other cases, what happens is a decrease in performance due to a reduction in the water-bearing thickness that is captured, forcing a decision to re-drill the borehole, if this is technically feasible. Other cases have been described in which the characteristics at depth become extremely low yielding, due to the almost total absence of karstification [55] in Chap. 1 [12].

Alteration of fluvial regime—In semi-arid karst regions, rivers tend to have no flow outside of high-intensity rain events. However, in many other regions there are perennial watercourses in karst areas, and intensive pumping has a corresponding effect, with notable environmental degradation in many cases [56]. This is the case in the Parks of Las Lagunas de Ruidera and Las Tablas de Daimiel, described as ecological catastrophes. Uncontrolled proliferation of borehole extraction has dried up some unique resources [57–59]. Moreover, in certain cases, it may be appropriate to arrange seasonal feeding from rivers to assure a water supply at times of low waters, or to carry out transfers, as in Las Tablas de Daimiel.

Changes in wetlands—In wetlands associated with karst aquifers, exploitation necessarily affects the ecology [60]. This is the case of the Ruidera lagoons in the province of Ciudad Real; 35 hm^3/year were pumped [61] from an aquifer of 2700 km^2 with an average annual recharge of 126 hm^3/year. Despite the extraction being on a much smaller scale than the average input, numerous lagoons dried up and the rest were considerably reduced in depth [57].

Legal complications with third parties—Problems arise over the acquired rights of spring water. The reduction in summer flow coincides with greater demand for water, which can bring pumping by traditional users to a standstill if their rights are not promoted. In many countries there have been attempts to regulate karst springs after detailed studies and costly exploitation boreholes. Deifontes in Granada, with a pumping capacity of 2.35 m³/s, and Pego are two instances of works that never been operational due to opposition from the existing users. Legal complications can lead to lengthy and costly proceedings aimed at protecting users' acquired rights.

8.5.2 Indirect Impacts

Salinization of soils—Irrigation with karst waters with a certain saline content can contribute to an increase in the salinity of the land. There are numerous examples where the salinity of the water is combined with high evaporation in a semi-arid environment. In other cases, phytotoxic elements from farming can be mobilized. Such is the case of the Deep Aquifer, made up of dolomites and marble dolomites from Andarax (Almeria), which contains several mg/l of boron, forcing plots and boreholes [62] to be abandoned. Moreover, Quaternary detrital material provides numerous salts to complicate the problem.

Progressive desertification—Plots abandoned by salinization are particularly vulnerable to erosion. In semi-arid regions of southeast Spain, parts of Morocco, Algeria and Tunisia the agricultural practices involved in cultivating citrus, among other fruit trees, readily promote erosion of the terraces, giving rise to *piping*, ravines and similar processes [63].

Collapse induction—Around the world there are many examples of karst collapse in which exploitation of the system has had a certain influence [50]. Water acts as a stabilizing element as it supports part of the load. Sudden drops in water level generate great instabilities and can lead to collapse. In other cases, pumping causes the mobilization of the clay that partially fills karst ducts, and the consequent imbalance can also promote collapse [64].

The countries with the most collapses are probably China and the United States [56]. The average dimensions of a collapse are less than 20 m across and less than 10 m in depth. The shape is variable, from circular to elliptical or elongated. In southern China, there have been more than 3000 pump-induced collapses since 1960. As a result of a collapse, buildings, roads, railways, pipelines and other infrastructure can be buried or destroyed. The damage caused by the collapse in Winter Park City (Florida) in May 1981 cost possibly more than US$4 million [65, 66].

When gypsum is involved, the process tends to take place faster as a result of the greater solubility [67]. Collapse and subsidence of saline diapirs (salt domes) and other evaporite accumulations are common, although not necessarily related to pumping. This is the case of the abandoned potash mines of Cardona in Catalonia, or the traditional salt mines of Wilyzca (Poland; Photo 8.21) that have had multiple subsidence problems.

Changes in the physical properties of the terrain—We start from the principle that karst and karstification encompass a series of processes in continuous activity, which translates into continuous change to the medium on the geological scale in

Photo 8.21 Deformed rails and broken walls as a result of settlement in the vicinity of the Wilyzca mines (Poland) (Photo A. Pulido)

Chap. 3 [6]. The major extraction points can converge and increase the potential for karstification, which will be promoted if, in addition, the water mixture produced causes subsaturation in calcite (karstification by waters mixing in Chap. 2 [3]. Islands are a particular case [68], since the mixing of freshwater and seawater usually makes for subsaturated water with respect to calcite, resulting in considerably increased karstification. This is especially evident across large areas of Cuba (Photos 8.22 and 8.23).

Contamination induced at distance, due to intensive pumping—The conoid produced by intensive pumping can mobilize polluted water at a distance and thus extract it, with consequent damage. This is the case of water polluted by discharge from a paper mill into a tributary of the River Vistula near the Olkusz lead mines, which had to be intensively drained to enable the extraction to proceed [54]. As it involved pollutants of an orogenic nature, the risk of *firedamp* increased considerably.

The mobilization of pollutants can also be achieved by the discharge of toxic waste into deep formations in carbonate aquifers, very common practice in Florida, for example. The decompression from the pumping can induce an increase in contamination through the fractures and by exploiting any imperfections in the injection boreholes [66].

Photo 8.22 *Cashimba* (freshwater thermal pools in coastal dolinas) in the area of Caleta Buena, Cuba (Photo A. Pulido)

Photo 8.23 Another cashimba in Caleta Buena, Cuba (Photo A. Pulido)

8.6   Further Reading

Alcalá, F. J., Custodio, E. 2008. Using the Cl/Br ratio as a tracer to identify the origin of salinity in aquifers in Spain and Portugal. *Journal of Hydrology, 359*, 189–207.

Colombani, N., Osti, A., Volta. G., Ma Gorelick, S. M., Zheng, C. 2015. Global change and the groundwater management challenge. *Water Resources Research, 51*(5), 3031–3051.

Goldscheider, N. 2012. A holistic approach to groundwater protection and ecosystem services in karst terrains. *AQUA Mundi*–Am06046, 117–124.

Bonacci, O., Pipan, T., Culver, D. C. 2009. A framework for karst ecohydrology. *Environmental Geology, 56*, 891–900.

Gaspar, E. (ed.). 1987. *Modern Trends in Tracer Hydrology.* Boca Raton: CRC Press, vol. 2.

Goldscheider, N., Drew, D. 2007. *Methods in Karst Hydrogeology.* London: Taylor and Francis.

Hötzl, H., Werner, A. (eds.). 1992. *Tracer Hydrology.* Rotterdam: Balkema, p. 464.

Li, S. C., Wu, J. 2018. A multi-factor comprehensive risk assessment method of karst tunnels and its engineering application. *Bulletin of Engineering Geology and the Environment.* https://doi.org/10.1007/s10064-017-1214-1

Ivan, V., Mádl-Szőnyi. 2017. State of the art of karst vulnerability assessment: Overview, evaluation and outlook. *Environmental Earth Sciences, 76*, 112.

Keller, E. A. 2010. *Environmental Geology*, 9th edn. Prentice Hall, p. 624.

Lauber, U., Ufrecht, W., Goldscheider, N. 2014. Spatially resolved information on karst conduit flow from in-cave dye tracing. *Hydrology and Earth System Sciences, 18*, 435–445.

Llamas, M. R., Custodio, E. 2010. *Intensive Use of Groundwater: Challenges and opportunities.* Rotterdam: Balkema, p. 481.

Mohammadi, Z., Gharaat, M. J., Field, M. 2019. The effect of hydraulic gradient and pattern of conduit systems on tracing tests: Bench-scale modeling. *Groundwater, 57*(1), 110–125.

Scanlon, B. R., Healy, R. W., Cook, P. G. 2002. Choosing appropriate techniques for quantifying groundwater recharge. *Hydrogeology Journal, 10*, 18–39.

Malard, A., Sinreich, M., Jeannin, P. -Y. 2016. A novel approach for estimating karst groundwater recharge in mountainous regions and its application in Switzerland. *Hydrological Processes, 30*(13), 2153–2166.

Fiorillo, F., Pagnozzi, M., Ventafridda, G. 2015. A model to simulate recharge processes of karst massifs. *Hydrological Processes, 29*, 2301–2314.

8.7   Short Questions

1. List the main exploration techniques for karst aquifers.
2. Which geophysical prospecting methods are most effective in a karst environment?
3. What conditions need to be met to resolve isothermal and isoconductivity curve profiles and maps? Give reasons for your answer.
4. List the main methods for estimating recharge in karst aquifers.
5. Why does quantification of a karst aquifer's recharge often have many uncertainties? Give reasons for your answer.
6. List the main objectives that can be met by the use of tracers in karst.
7. What special precautions should be taken when using radioactive tracers in a karst environment?
8. Are karst aquifers easy to regulate and tap?
9. List the main methodologies for regulating and tapping a karst spring.

10. Is using galleries a sustainable technique for the regulation of a karst aquifer? Why is this?
11. Which drilling technique is fastest and most conclusive in a karst environment?
12. Is it is recommended to drill boreholes by means of direct rotation with bentonite sludge?
13. List the main steps in performing acidification, and the safety measures to be taken.
14. Why is it so difficult to tap coastal and/or submarine springs?
15. List the main effects of the exploitation of karst aquifers.

8.8   Personal Work

1. Using tracers in a karst medium.
2. Rotor percussion drilling in karst aquifers: problems and solutions.
3. Tapping galleries in karst aquifers: describe several exemplars.
4. Tapping submarine springs: describe several exemplars.
5. The overexploitation of karst aquifers.

References

1. Bechtel, T. D., Bosch, F. P., & Gurk, M. (2007). Geophysical methods. In N. Goldscheider, & D. Drew, (eds.), *Methods in karst hydrogeology*. Taylor and Francis, London, pp. 171–199.
2. Potié, L., & Tardieu, B. (1977). Aménagement et captage sous-marins dans les formations calcaires. *Karst Hydrogeology* (Congreso de Alabama, 1975), pp. 39–56.
3. Potié, L. (1989). *La résurgence d'eau douce sous-marine de Port-Miou* (p. 104). Chantiers de France.
4. Gaspar, E. (Ed.). (1987). *Modern Trends in Tracer Hydrology*. (Vol. 2). Boca Raton: CRC Press.
5. Hötzl, H., & Werner, A. (Eds.). (1992). *Tracer Hydrology* (p. 464). Rotterdam: Balkema.
6. Benischke, R., Goldscheider, N., & Smart, C. (2007). Tracer Techniques. In N. Goldscheider & D. Drew (Eds.), *Methods in karst hydrogeology* (pp. 147–170). London: Taylor and Francis.
7. Lepiller, M., & Mondain, P. H. (1988). Les traçages artificiels en hydrogéologie karstique. *Hydrogéologie, 1*, 33–52.
8. Delay, F., & Bracq, P. (1993). Analysis of the spatial distribution of morphological features applied to the needs of hydrogeology. *Computers and Sciences, 19*(7), 965–980.
9. Doerfliger, N., Zwahlen, F. (1996). EPIK: a new method for the delineation of protection areas in karstic environment. In I. Johnson, G. Gunay, (Eds.) *Karst water and human impacts*, Balkema.
10. Daniele, L., Pulido-Bosch, A., Vallejos, A., & Molina, L. (2008). Geostatistical analysis to identify hydrogeochemical processes in complex aquifers: A case study (Aguadulce unit, Almeria, SE Spain). *Ambio, 37*(4), 249–253.
11. Granda, A. (1986). La geofísica aplicada en hidrogeología. Algo más que el SEV. *Boletin Geologico y Minero, XCVII*(I), 65–76.
12. Tulipano, L., & Fidelibus, M. D. (1995). National report from Italy. In *Karst groundwater protection*. EUR 16547 EN, pp. 171–201.

13. Molina Sanchez, L. (1998). *Hidroquímica e intrusión marina en el Campo de Dalías (Almería)* (340 p.). Granada: Tesis Doct. University of Granada.
14. Kiraly, L., et al. (1995). Numerical simulation. In *National Report from Switzerland. Karst groundwater protection*. Final Report EUR 16457 EN, pp. 279–304. Bruselas.
15. Bakalowicz, M., et al. (1994). High discharge pumping in a vertical cave. Fundamental and applied results". In Crampon, N., & Bakalowicz, M. (eds.), *Hydrogeological aspects of groundwater protection in karstic areas* (pp. 93–110). E.C.D.G. Bruselas.
16. Tóth, J. (1999). Groundwater as a geologic agent: An overview of the causes, processes, and manifestations. *Hydrogeology Journal, 7*, 1–14.
17. Bakalowicz, M. (1996). Impacts of pumping in karst aquifers. In Hötzl, H., & Drew, D. (Eds.), *Impacts of human activities on karstic hydrogeology*, p. IAH book, n°17.
18. Sheneveli, L. (1996). Analysis of well hydrographs in a karst aquifer: estimates of specific yields and continuun transmissivities. *Journal of Hydrology, 174*, 331–355.
19. Rouhiainen, P. (1993). A flowmeter for groundwater in fractured bedrock. *XXVI IAH Congress, 2*, 762–771. Oslo.
20. Lerner, D. N., Issar, A. S., & Simmers, I. (1990). Groundwater recharge. A guide to understanding and estimating natural recharge. *International Contributions to Hydrogeology, 8*, 345. IAH–Heise, Hannover.
21. Alcalá, F. J., Cantón, Y., Contreras, S., Were, A., Serrano-Ortiz, P., Puigdefábregas, J., et al. (2011). Diffuse and concentrated recharge evaluation using physical and tracer techniques: Results from a semiarid carbonate massif aquifer in southeastern Spain. *Environmental Earth Sciences, 62*(3), 541–557.
22. Andreu, J. M., Alcalá, F. J., Vallejos, A., & Pulido-Bosch, A. (2011). Recharge to mountainous carbonated aquifers in SE Spain: Different approaches and new challenges. *Journal of Arid Environments, 75*, 1262–1270.
23. Contreras, S., Boer, M. M., Alcalá, F. J., Domingo, F., García, M., Pulido-Bosch, A., et al. (2008). An ecohydrological modelling approach for assessing long–term recharge rates in semiarid karstic landscapes. *Journal of Hydrology, 351*, 42–57.
24. Nemani, R. R., & Running, S. W. (1989). Testing a theoretical climate-soil-leaf area hydrologic equilibrium of forests using satellite data and ecosystem simulation. *Agricultural and Forest Meteorology, 44*, 245–260.
25. Specht, R. L., & Specht, A. (1989). Canopy structure in Eucalyptus-dominated communities in Australia along climatic gradients. *Oecologica Plantarum, 10*, 191–202.
26. Boer, M., & Puigdefábregas, J. (2003). Predicting potential vegetation index values as a reference for the assessment and monitoring of dryland condition. *International Journal of Remote Sensing, 24*, 1135–1141.
27. Contreras, S., (2006). *Spatial distribution of the annual water balance in semiarid mountainous regions: Application to Sierra de Gádor (Almería, SE Spain)*. Thesis Doct., University of Almeria, Spain.
28. Custodio, E., Llamas, M. R., & Samper, J. (Eds.). (1997). *La evaluación de la recarga a los acuíferos en la planificación hidrológica* (p. 453). Madrid: IAH-GE and ITGE.
29. Scanlon, B. R., Healy, R. W., & Cook, P. G. (2002). Choosing appropriate techniques for quantifying groundwater recharge. *Hydrogeology Journal, 10*, 18–39.
30. Alcalá, F. J., & Custodio, E., (2008). Atmospheric chloride deposition in continental Spain. *Hydrological Processes, 22*, 3636–3650.
31. Alcalá, F. J., & Custodio, E. (2008). Using the Cl/Br ratio as a tracer to identify the origin of salinity in aquifers in Spain and Portugal. *Journal of Hydrology, 359*, 189–207.
32. Aguilera, H., & Murillo, J.M., (2009). The effect of possible climate change on natural groundwater recharge based on a simple model: a study of four karstic aquifers in SE Spain. *Environmental Geology, 57*, 963–974.
33. Fidelibus, M. D., & Tulipano, L. (1990). Major and minor ions as natural tracers in mixing phenomena in coastal carbonate aquifers of Apulia. In *11th SWIM*, (pp. 283–293). Gdansk.

34. Goldscheider, N., Meiman, J., Pronk, M., & Smart, C. (2008). Tracer tests in karst hydrogeology and speleology. *International Journal of Speleology, 37*(1), 27–40.
35. Machkova, M. Vallejos, A., Pulido-Bosch, A., Dimitrov, D., Calaforra, J. M., & Gisbert, J. (2002). Investigation of mountain karst systems behaviour by tracer techniques (on the example of the Nastan-Trigrad karst system-Bulgaria). *XXIst Conference of the Danubian Countries.* Bucarest, Rumania, pp. 1–9.
36. Pulido-Bosch, A., Litchev, A., Machkova, M., Dimitrov, D., López Chicano, M., Calvache, M. L., Calaforra, J. M., Velikov, B., & Pulido Leboeuf, P. (2002). Aplicación de técnicas geofísicas e isotópicas para determinar variaciones verticales de parámetros hidráulicos en acuíferos heterogéneos (ejemplo de Dobrich, NE de Bulgaria). *3ª Asamblea Hispano-Portuguesa de Geodesia y Geofísica,* Vol. III, pp. 1836–1839.
37. Pulido-Bosch, A., López Chicano, M., Machkova, M., Dimitrov, D., Velikov, B., Calaforra, J. M., & Calvache, M. L. (1999). Karst water environmental problems at the town of Dobrich, NE Bulgaria. *Groundwater in the urban environment: Selected city profiles* (pp. 225–231), Rotterdam.
38. Plata, A., (1983). Single well techniques using radioactive tracers. *Tracer Methods in Isotope Hydrology.* IAEA–TECDOC-291, 17–33. Vienna.
39. Plata, A. (1991). Detection of leaks from reservoirs and lakes. *Use of artificial tracers in hydrology,* IAEA, 71–129. Vienna.
40. Eisenlohr, L. (1995). *Variabilité des réponses naturelles des aquifères karstiques.* Neuchâtel: Tesis Doct. Univ.
41. Pavlin, B., & Fritz, F. (1978). La protection du système des sources karstiques de Golubinka contre la contamination par la mer. *SIAMOS, 1,* 227–235, Granada.
42. Mijatovic, B. F. (1984). Captage par galerie dans un aquifère karstique de la côte dalmate–Rimski Bunar, Trogir (Yougoslavie). In A. Burger, & L. Dubertret (Eds.), *Hydrogeology of karstic terrains. Case Histories.* IAH, Vol. 1, pp. 152–155.
43. Mijatovic, B. F. (Ed.). (1984). *Hydrogeology of the Dinaric karst* (Vol. 4, p. 255). Heise: IAH.
44. Pulido-Bosch, A., Castillo, A., & Padilla, A. (Eds.). (1989). *La sobreexplotación de acuíferos.* Temas Geológico-Mineros, (Vol. 10, p. 687). Madrid: IGME.
45. Candela, L., Gómez, M. B., Puga, L., & Rebollo, L., Villarroya, F. (1991). *Aquifer overexploitation, XXIII,* (p. 580). IAH Congress: Canarias.
46. Custodio, E. (2002). Aquifer overexploitation: What does it mean? *Hydrogeology Journal, 10,* 254–277.
47. Llamas, M. R., & Custodio, E. (2002). Acuíferos explotados intensivamente: conceptos principales, hechos relevantes y algunas sugerencias. *Boletín Geológico y Minero, 113*(3), 223–228.
48. Llamas, M. R., & Custodio, E. (2002). *Intensive Use of Groundwater: Challenges and Opportunities* (p. 481). Balkema.
49. Simmers, I., Villarroya, F., & Rebollo, L. F. (Eds.). (1992). *Selected papers on aquifer overexploitation.* IAH Selected Papers,(Vol. 3, p. 391). Hannover: Heise.
50. Lamoreaux, P. E. (1991). *Environmental effects to overexploitation in karst terranes. XXIII IAH Congress* (Vol. I, pp. 103–113). Canarias.
51. Rodríguez Estrella, T., & Gómez de las Heras, J. (1986). Principales características de los acuíferos kársticos de la província de Murcia. In *Jornadas sobre el Karst en Euskadi* (Vol. 1, pp. 187–203). San Sebastián.
52. Pulido-Bosch, A. (Ed.). (1993). *Some Spanish karstic aquifers* (p. 310). University of Granada.
53. Calvache, M.L., Pulido-Bosch, A. (1994). Modelling the effects of salt–water intrusion dynamics for a coastal karstified block connected to a detrital aquifer. *Ground Water, 32,* 767–777.

54. Motyka, J., Witzak, S., & Zuber, A. (1994). Migration of lignosulfonates in a karstic fractured-porous aquifer; History and prognosis for a Zn–Pb mine, Pomorzany, southern Poland. *Environmental Geology, 24,* 144–149.
55. LeGrand, H. E., & Stringfield, V. T. (1971). Development and distribution of permeability in carbonate aquifers. *Water Resources Research, 7,* 1284–1294.
56. Volker, A., & Henry, J. C., (Eds). (1988). Side effects of water resources management. *International Association of Hydrological Sciences Publication, 172,* 269.
57. Montero, E. (1994). *Funcionamiento hidrogeológico del sistema de las Lagunas de Ruidera* (p. 297). Tesis Doct., University Complutense, Madrid.
58. Llamas, R., Back, W., & Margat, J. (1992). Groundwater use: Equilibrium between social benefits and potential environmental costs. *Applied Hydrogeology, 1,* 3–14.
59. Martínez Alfaro, P., Montero, E., & López Camacho, B. (1992). The impact of the overexploitation of the Campo de Montiel aquifer on the Lagunas de Ruidera ecosystem. *Selected Papers on Hydrogeology, 3,* 87–91.
60. Llamas, M. R., & Martínez-Santos, P. (2005). Intensive groundwater use: Silent revolution and potential source of social Conflicts. *Journal of Water Resources Planning and Management,* 337–341.
61. DGOH, DGCA, ITGE, (1994). *Libro blanco de las aguas subterráneas* (p. 135). Madrid: MOPTMA
62. Pulido-Bosch, A., Sánchez Martos, F., Martínez Vidal, J. L., & Navarrete, F. (1992). Groundwater problems in a semiarid area (Low Andarax River, Almeria, Spain). *Environmental Geology and Water Sciences, 20*(3), 195–204.
63. García Ruíz, F. J., Lasanta, T., Ortigosa, L., & Amaez, J. (1986). Pipes in cultivated soils of La Rioja: Origin and evolution. *Zeitschrift für Geomorphologie. Supplementband, 58,* 93–100.
64. Garay, P. (1986). Informe geológico sobre la sima de hundimiento de Pedreguer (Alicante). *Jornadas Karst Euskadi, 1,* 323–333.
65. Dougherty, P. H., & Perlow, M. (1987). The Macungie sinkhole, Lehigh valley, Pensilvania: Cause and repair. *Environmental Geology and Water Science, 12*(2), 89–98.
66. Keller, E. A. (2010). *Environmental Geology* (9th edición, p. 624). Prentice Hall.
67. Galve, J. P., Gutiérrez, F., Remondo, J., Bonachea, J., Lucha, P., & Cendrero, A. (2006). Evaluating and comparing methods of sinkhole susceptibility mapping in the Ebro valley evaporite karst (NE Spain). *Geomorphology, 111*(3), 160–172.
68. Back W., (1992). Coastal karst formed by ground–water discharge, Yucatan, México. *International Contributions to Hydrology, 13,* 461–466.